Manual of Biobank Quality Management

Agnieszka Matera-Witkiewicz •
Joanna Gleńska-Olender •
Izabela Uhrynowska-Tyszkiewicz •
Małgorzata Witoń • Karolina Zagórska •
Katarzyna Ferdyn • Michał Laskowski •
Patrycja Sitek • Błażej Marciniak •
Jakub Pawlikowski • Dominik Strapagiel

Manual of Biobank Quality Management

BBMRI.pl
Biobanking and
BioMolecular resources
Research Infrastructure
Poland

Springer

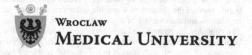

WROCLAW
MEDICAL UNIVERSITY

Agnieszka Matera-Witkiewicz
Faculty of Pharmacy, Wroclaw Medical
University Biobank; Screening of Biological
Activity Assays and Collection of Biological
Material Laboratory
Wroclaw Medical University
Wroclaw, Poland

Joanna Gleńska-Olender
Faculty of Pharmacy, Wroclaw Medical University
Biobank
Wroclaw Medical University
Wroclaw, Poland

Izabela Uhrynowska-Tyszkiewicz
Department of Transplantology and Central
Tissue Bank
Medical University of Warsaw
Warsaw, Poland

Małgorzata Witoń
Faculty of Pharmacy, Wroclaw Medical University
Biobank
Wroclaw Medical University
Wroclaw, Poland

Karolina Zagórska
Faculty of Pharmacy, Wroclaw Medical
University Biobank
Wroclaw Medical University
Wroclaw, Poland

Katarzyna Ferdyn
Faculty of Pharmacy, Wroclaw Medical University
Biobank
Wroclaw Medical University
Wroclaw, Poland

Michał Laskowski
Faculty of Pharmacy, Wroclaw Medical
University Biobank; Screening of Biological
Activity Assays and Collection of Biological
Material Laboratory
Wroclaw Medical University
Wroclaw, Poland

Patrycja Sitek
Faculty of Pharmacy, Wroclaw Medical University
Biobank
Wroclaw Medical University
Wroclaw, Poland

Błażej Marciniak
Faculty of Biology and Environmental
Protection, Biobank Laboratory, Department of
Oncobiology and Epigenetics
University of Lodz
Lodz, Poland

Jakub Pawlikowski
Department of Humanities and Social Medicine
Medical University of Lublin
Lublin, Poland

Dominik Strapagiel
Faculty of Biology and Environmental
Protection, Biobank Laboratory, Department of
Oncobiology and Epigenetics
University of Lodz
Lodz, Poland

ISBN 978-3-031-12561-4   ISBN 978-3-031-12559-1   (eBook)
https://doi.org/10.1007/978-3-031-12559-1

1st Edition © Wroclaw Medical University 2019 / 2nd Edition © Wroclaw Medical University 2021

This Springer imprint is published by the registered company Springer Nature Switzerland AG
The registered company address is: Gewerbestrasse 11, 6330 Cham, Switzerland

Preface

The importance of quality in every area of life is not needed to be convinced. Nowadays, a great attention is paid whether partner or institution meets the quality standards. Having to choose between a partner with implemented certain standards and one without it, the choice is usually simple and clear. Do we pay the same attention to the scientific partners? Are we reviewing the quality standards of their work? Implementing and maintaining quality standards of the biological material on which we plan to base our project should also be an obviousness. But is it?

Most Biobanks themselves establish internal rules of their operation, which in most cases are the rules of the organizations in which they operate. During scientific cooperation between Biobanks, many problems may arise related to the lack of uniform rules for handling biological material and data.

Due to the lack of a single guide, describing the most important aspects of biobanking, the idea for the first version of this manuscript was born. The first version has been published only in the polish language as *"Quality Standards for Polish Biobanks,"* but after extensive reviewers within the biobanking community the authors decided to prepare another manual based on the first idea of the book. The work you are reading *"Manual of Biobank Quality Management"* is the second updated and completed edition intended to be not only a set of guidelines resulting from the normative regulations, but also a source of practical knowledge, a friendly reading guide—which can be referred to the organization of a new Biobank as well as during everyday work and self-improvement.

The *"Manual of Biobank Quality Management"* is divided into 15 chapters. Each of them includes the main part containing a substantive contribution referring strictly to the requirements for Biobanks, including the ISO 9001:2015 and ISO 20387:2018 along with an indication of references to individual standard points. It also contains a practical section covering: *the most common practices* in Biobanks and *Frequently Asked Questions*—FAQs. The most commonly used practices section is based on advice, tips, and even ready-made solutions that are used in Biobanks.

It has resulted from expertise knowledge in the field of biobanking, tissue and cell banking, laboratory diagnostics, biology and biotechnology supported by significant practical experience. Also the specialists in the field of ELSI (ethical, legal, and societal issues) and IT have been invited to cooperate with interdisciplinary approach and completion.

The manual presents only proven methods that guarantee efficient Biobank management and organization while minimizing the risk and nonconforming processes outputs. The FAQ section contains questions we encountered during trainings, audits, webinars, and conferences. It is the result of plenty of hours of discussions between the authors of the handbook and Biobank's employees as well as of detailed analyses of satisfaction surveys conducted after the trainings and webinars. Thanks to this solution, our readers will have the opportunity to solve the most common doubts bothering them on their way to meeting the requirements.

The first chapter *Management of Biobanks* describes the requirements to be fulfilled by the Biobank concerning its organizational structure, mission and scope of activities, setting and monitoring of objectives, as well as internal and external communication processes. The purpose of communication is to ensure an effective information flow between the organizational units of the organization/unit in which the Biobank operates; within the Biobank itself and between the Biobank and stakeholders.

Chapter two *Quality Management* describes how Biobank shall establish the basic assumptions and principles of the QMS, by taking the process approach in Biobank management; ensuring continuous QMS improvement, supervision over the processes, and the resources necessary for their functioning; constant focus on the needs and requirements of interested parties, including research participants, clients, and partners. Draws attention to the need to develop standard operating procedures covering management, operational, and technological processes to ensure that the implemented QMS is effective. This chapter also describes how the risk and opportunity shall be identified, evaluated, analyzed, and monitored as well.

Chapter three, *Documentation and Records*, concerns internal documentation where main guidelines regarding documentation creation and ensuring its traceability, readability, availability, confidentiality, control of changes, back-up documentation, its implementation, updating and withdrawal, distribution, storage, and verification are indicated.

Chapter four *Human Resources Management* describes the human resource management policy, the process of employee identification, the hiring process (including the scope of initial training), on-the-job trainings, and verification of employees' knowledge as well. Additionally, it describes how to establish responsibility and authority of biological material biobanking process when cooperation with the staff is terminated.

Chapter five, *Ethical, Legal, and Societal Aspects,* is dedicated to compliance with general regulations and norms such as impartiality, confidentiality, and respect for autonomy including informed consent, sharing biological material and data, proprietary rights and commercialization.

Chapter six, *Supplies, Materials Management*, describes the management of materials and reagents, including their receipt, qualification, and acceptance for use, the process of qualifying suppliers, the handling of materials stored in the warehouse, and the handling of materials used in certain operations.

In the chapter seven *Equipment,* the principles for the acceptance of equipment and requirements for installation, operational and performance qualification process are profiled. It indicates the principles of internal and external devices supervision. The handling of devices outside supervision and the process of reclassification of key devices are pointed.

Chapter eight *Traceability* depicts a system ensuring traceability of biological material and related data received into and issued from Biobank. It also presents the unique labeling of biological material and related data, making it possible to be identified at each stage of their life cycle.

In the chapter nine, *Environmental and Staff Hygiene,* occupational health and safety, occupational risk, handling of protective clothing and equipment, work with potentially infectious material, occupational exposure, processes environment, including room hygiene, and waste management as one of the critical issues are presented.

Chapter tenth *Biobanking Processes and Quality Control* describes the life cycle of biological material and critical technological processes, including the acquisition/ collection of biological material and related data, transportation, reception, qualification, and acceptance of biological material and related data, processing and storage as well as quality control process requirements.

Chapter eleventh *Deviations, Nonconforming Product/Data or Service* describes the deviations and nonconformities identified in Biobank; corrective actions and complaints. In addition, it describes the process of identifying nonconforming products/services and rejecting nonconforming outputs.

In chapter twelve *Audits,* the process of conducting internal audits at Biobank, creating the audit program, preparation audits phase, conducting the audit and post-audit activities is carefully prepared. Additionally, this section describes the competences and qualifications of auditors as well.

Chapter thirteen, *Improvement,* describes the tools for biobanking system improvement. This section describes how to plan, conduct, and document a management review aimed at ensuring the suitability, adequacy, and effectiveness of QMS and processes in Biobank.

Chapter fourteenth *Scientific Cooperation* describes the process of sharing biological material and associated data for research purposes. Moreover, the guidelines for concluding cooperation agreements are also presented. It raises the issue of communication during scientific cooperation, including information about irregularities in the course of research.

And the last but not least, Chapter fifteen *Safety and Security* is dedicated to safety procedures, including the safety of biological material and information security. This point highlights the basic methods of securing IT infrastructure and data processing, including the awareness of the Biobank functioning in the organization, general principles of secure information exchange, security of processed data, basic methods of securing access to office and laboratory space, operating systems and software, basic features of the Biobank IT system, backup copies or external services, including data processing in the cloud.

All requirements and instructions are guided by the principles of responsibility, openness, and transparency.

The Manual of Biobank Quality Management is friendly in form and content guide for anyone who is just starting their adventure with biobanking or has extensive experience in this field and only wants to complete their knowledge. Close correlation with ISO 9001:2015 and ISO 20387: 2018 standards constitutes a very good basis for the creation and efficient management of a Biobank, as well as preparation for emerging situations and problems. The manual pointed out that not only the quality of biological material is important, but also the quality of associated data and information regarding to them.

It is great pleasure and honor that we can hand over our work, hoping that it will be a useful and inspiring adventure on the road to improving your biobanking processes and preparing for accreditation to confirm the competences or certification process to prove the quality.

Wroclaw, Poland   Agnieszka Matera-Witkiewicz
Wroclaw, Poland   Joanna Gleńska-Olender
Warsaw, Poland   Izabela Uhrynowska-Tyszkiewicz
Wroclaw, Poland   Małgorzata Witoń
Wroclaw, Poland   Karolina Zagórska
Wroclaw, Poland   Katarzyna Ferdyn
Wroclaw, Poland   Michał Laskowski
Wroclaw, Poland   Patrycja Sitek
Lodz, Poland   Błażej Marciniak
Lublin, Poland   Jakub Pawlikowski
Lodz, Poland   Dominik Strapagiel

Acknowledgments

BBMRI.pl, project founded by the Ministry of Science and Higher Education DIR/WK/2017/2018/01-1.

Contents

Abbreviations

BBMRI	Biobanking and Biomolecular Resources Research Infrastructure
BM	Biological Material
IEC TR	International Electrotechnical Commission Technical Report
IQ	Installation Qualification
ISO	International Organization for Standardization
MBQM	Manual of Biobank Quality Management
OQ	Operational Qualification
PQ	Performance Qualification
REC/BC	Research Ethics Committee/Bioethic Committee
QMS	Quality Management System
TS	Technical Specification

Management of Biobanks

1

Abstract

The Biobank management is a process that orientates the organization toward achieving the objectives set in an efficient and effective manner. Efficiency means a favorable ratio of inputs to outputs. Effectiveness, on the other hand, means that all planned activities are performed and, thus, the objectives assumed are accomplished. One of the key roles in ensuring greater compatibility between the organization and its environment is its structure, i.e., the arrangement of the units of the organization and the interrelationships between them. The specificity of Biobank management manifests itself primarily in the scope of its activities, mission, objectives, specified environment, structures, and key values. To ensure the proper functioning of the Biobank organization and smooth cooperation between people, the management should make every effort to establish the principles of internal and external communication process, including, among other things, the flow of information, the principles of issuing documents inside and outside Biobank as well as the principles of communication with external organizations and stakeholders (including research participants).

- A Biobank shall be a legal entity or a specific part of a legal entity.
- The Biobank shall conduct its operations in such a way as to meet the requirements of superior documentation governing its work (statutes, act of appointment, resolution) and obligations arising from the concluded cooperation agreements. Biobank's work shall be subject to internal and external supervisory mechanisms (see Sect. 5.2 MBQM)
- Organization Management—the Biobank shall include a Quality Management System (see Chap. 2 MBQM)

5.1, 5.6, 5.7, 5.9, 8.1.1, ISO 20387:2018
4.3, 4.4.1 ISO 9001:2015

© Wroclaw Medical University 2023
A. Matera-Witkiewicz et al., *Manual of Biobank Quality Management*,
https://doi.org/10.1007/978-3-031-12559-1_1

1.1   Organizational Structure

1.1.1   Organizational Structure of the Organization/Institution Under Which the Biobank Operates

- The institution within which the Biobank operates shall have an established and approved organizational structure.
- The Biobank shall plan, implement, and supervise processes related to meeting the requirements imposed by the organization (e.g., based on the status of the organization together with documents regulating the organizational structure and method of operation).

5.8 a) ISO 20387:2018

The Most Common Practices
- The organizational structure of the Biobank may vary depending on the nature of the unit and the resources it has. Its reflection and extension should be the Organizational Regulations of the Unit.
- The Biobank's organizational structure should be available to interested parties (e.g., placed on the website of the organization/institution).
- For the designing of a new scientific study, reference should be made to the organizational structure, i.e., to verify which departments are affected by the design and who will be responsible for the approval of individual stages.

1.1.2   Organizational Chart of the Biobank

- Biobank shall have an approved organizational chart, in which the relationships between organizational/functional units are specified.
- The organizational chart of the Biobank shall include the unit/person responsible for the Quality Management System.
- Biobank shall be subordinate to the Scientific Committee/Scientific Council (if any) or other advisory body in the field of scientific, technical, and/or administrative projects to which the Biobank is subject to.

5.3, 5.8 ISO 20387:2018

The Most Common Practices
- In the organizational chart of the Biobank, it is recommended to enter the names of individual job positions.
- The structure may be graphic and apply to the Biobank itself, and the subordination to the advisory body may be described in the procedure or may result from general guidelines/ordinances/regulations of the organization (e.g., university statutes).

Frequently Asked Questions
- **Should the organization chart be updated after the admission of each new employee?**
  Yes, after the admission of each new employee, the organizational chart of the Biobank should be updated, and then each employee should become acquainted with it and sign it confirming this fact.

1.2   Biobank's Mission and Scope of Activity

- The Biobank shall specify the mission and scope of activity.
- The mission and scope of Biobank's activities shall be enforced and regularly reviewed.
- The Biobank shall aim to employ a sufficient number of qualified personnel and have rooms that allow for proper technological processes.

4.1.3, 5.5, 5.7, 5.9, 6.1.1, 8.2.2, 8.2.3, 8.9.1 ISO 20387:2018
5.1.1 b), 5.2.1, 5.2.2, 7.1, 9.1.1, 9.3.1 ISO 9001:2015

1.3   Biobank's Objectives and Their Monitoring

- Biobank shall implement the organizational policy, mission, and objectives. The assumed objectives shall be defined in time, measurable, updated if necessary and subject to analysis (see Sect. 2.3 MBQM).

4.1.3, 5.7, 5.9, 5.10, 8.2.2, 8.9.1 ISO 20387:2018
5.1.1 b), 5.2.1, 5.2.2, 7.1, 7.3, 7.5.1 b), 9.1.1, 9.3.1 ISO 9001:2015

The Most Common Practices
- Strategic objectives mainly include broad and general problems/issues. Strategic objectives are the basis for formulation operational objectives. Operational objectives are short-term and are primarily the responsibility of mid- or lower-level employees.
- It is a common practice that the Biobank verifies the set and approved goals at least once every 12 months.
- It is a common practice that Biobank sets and approves new goals for implementation at least once every 12 months.

1.4   Communication

- The Biobank shall conduct internal and external communication processes in such a way as to ensure an effective flow of information:
  - between individual organizational units of the organization/unit in which structures the Biobank functions and the Biobank;
  - within the Biobank;
  - between the Biobank and stakeholders (see Chap. 14 MBQM).
- The Biobank shall specify the procedure for dealing with complaints, adverse incidents or reactions (see Chap. 11 MBQM).
- The implemented communication process shall be known to Biobank employees.

  5.10 b), 5.10 c), 6.4.1.1 b), 6.4.1.3, 7.13.1, 8.2.1, 8.6.2, ISO 20387:2018)
  7.4, 8.2.1, 8.4.3, 10.2.1 ISO 9001:2015

1.4.1   External Communication

- The Biobank shall define the rules of communication with external organizations and stakeholders (including donors).
- These rules shall indicate who and how they can communicate and how it should be documented.

  5.10 b), 5.10 c), 6.4.1.1, 8.2.1, ISO 20387:2018
  7.4, 8.2.1, 8.4.3 ISO 9001:2015

The Most Common Practices
- In the case of commercial services provided by the Biobank, guidelines should be developed on what services are provided, costs of providing these services, available service hours, and contact information appropriate for each category of stakeholders during the assumed working hours, as well as in emergency situations.
- Biobank rules for submitting applications and complaints related to Biobank's work should be available to stakeholders.

1.4.2   Internal Communication

- The Biobank shall specify the way of information flow, issuing internal documents (ordinances, circulars, official orders) within and outside the Biobank.

  5.10 b), 5.10 c), 8.2.1, 8.2.6 ISO 20387:2018
  7.4, 8.2.1, 8.4.3 ISO 9001:2015

The Most Common Practices

- The communication process containing information on the circulation of originals and copies of documents and allowing documents to be scanned is presented in graphic form.

References

ISO 20387:2018 Biotechnology – Biobanking – General Requirements for Biobanking.
ISO 9001:2015 Quality management systems - Requirements.

Quality Management

2

Abstract

The quality management is an approach through which the Biobank is able to improve the effectiveness of all processes taking place in the organization in order to meet relevant requirements, needs, and expectations of customers. All activities may have an impact on the quality of biological material and/or associated data, that is why the quality management requires the cooperation between all employees of the Biobank. In the context of the quality management process, every staff member should make a particular effort to improve his or her own work. In this chapter, issues related to the quality policy, quality objectives and risk, and opportunity management will be discussed. The quality policy sets the direction in which the Biobank should develop its activities. Well-established objectives provide guidelines for employees. They can motivate them to work, facilitate the planning of activities, and ultimately improve the quality of work. They also make it possible to control and monitor the progress of activities. The risk and opportunity management allows the Biobank to properly manage its processes and, consequently, to effectively achieve its objectives.

2.1 General Requirements

- The quality management system (QMS) including quality assurance (QA) and quality control (QC) programs shall cover the full spectrum of Biobank operations.
- The Biobank shall establish the basic assumptions and rules of the Quality Management System:
  - application of a process approach in the management of the Biobank;
  - ensuring continuous improvement of the Quality Management System;

A. Matera-Witkiewicz et al., *Manual of Biobank Quality Management*,
https://doi.org/10.1007/978-3-031-12559-1_2

- ensuring supervision of the processes and necessary resources for their functioning;
- constant orientation to the needs and requirements of Biobank's stakeholders.
• The Biobank should establish standard operating procedures covering main and supporting processes to ensure that the implemented QMS is effective.

4.1.1, 6.1.2, 8.1.1, 8.1.2, 8.2 ISO 20387:2018
4.3., 4.4.1, 6.1.1, 8.1., 9.1.3 g), 10.1, 10.3 ISO 9001:2015

2.2   Quality policy

• The Biobank shall establish a quality policy.
• The quality policy shall be consistent with the objectives of Biobank's operation.
• It shall be understood by Biobank employees.
• It shall contain quality aspects relevant for the functioning of the Biobank.
• The quality policy shall be documented, implemented, and maintained.
• It shall be evaluated during Biobank's management reviews.

8.2.2, 8.2.3, 8.2.4., 8.9.1 ISO 20387:2018
5.2.1 a), 5.2.1 b), 5.2.2 b) ISO 9001:2015

The Most Common Practices
• The quality policy is often placed in visible places in the building of the unit. This helps to meet the requirements for transmission of quality policy content to Biobank's stakeholders.

Frequently Asked Questions
• **How often should the Biobank evaluate the Quality Policy?**
  The most common practice is to evaluate the Quality Policy once every 6 and 12 months. The Biobank internally sets the frequency of quality policy assessment in accordance with the Biobank's work schedule.

2.3   Setting Objectives

• Quality objectives shall be set at the Biobank, which should:
  - correspond with the quality policy of the Biobank;
  - be consistent and measurable;
  - take into account the quality objectives of samples and corresponding data;
  - be disseminated within the Biobank as a document signed by the Top Management;
  - be analyzed.

8.2.2, 8.2.3 ISO 20387:2018
5.2.1 a), 5.2.1 b), 5.2.2 b), 6.2.1 ISO 9001:2015

The Most Common Practices
- Examples of objectives for the Biobank:
  - supporting academic, industrial, and cooperative research influencing the development of biomedical sciences and improving diagnostics and therapy;
  - biobanking of biological material as part of cooperation with clinical hospital units;
  - biobanking of biological material as part of regional, domestic, and foreign research and programs;
  - ensuring the improvement of access to high standard biological material for units operating within the University;
  - ensuring the improvement of access to high standard biological material for external partners as part of scientific cooperation;
  - allowing the University's internal units to store accumulated biological material in stable, safe, and constantly monitored conditions;
  - applying for scientific research projects and programs in the field of biobanking of biological material and in the field of research using biological material.

Frequently Asked Questions
- **How often should the Biobank analyze its goals?**
  The most common practice: The Biobank should analyze its goals at least once every 12 months.

2.4 Risk and Opportunity Management

- The Biobank shall establish a risk and opportunity management procedure, which constitutes the basis for manage processes and, consequently, to enable the successful implementation of the objectives. Risk and opportunity assessment shall be carried out at specified intervals.
- The identification of risk and opportunity applies to all stages of biobanking of biological material, risk and opportunity, as well as their monitoring and annual reviews.

4.1.1, 6.3.7, 8.1.2 d), 8.5.2, 8.6.1, 8.9.2 k), m) ISO 20387:2018
6.1.1 a), 6.1.2 a) ISO 9001:2015

2.4.1 Identification of Risk and Opportunity

- The Biobank shall identify risks that should or may have an impact on the performance of Biobank's goals and internal operations, including the quality of biological material.

- While identifying risk and opportunity it is necessary to analyze the scope of operations of individual processes specified in the internal standard operation procedure and to analyze risk and opportunity related to their implementation and to define a corrective action plan.

8.5.1 ISO 20387:2018
6.1.2 ISO 9001:2015

The Most Common Practices
- Risk identification answers the following question: "what might go wrong?" and determines possible consequences.
- The following methods are used to identify a risk:
  - checklist,
  - brainstorming,
  - Ishikawa diagram,
  - FMEA *(Failure mode and effects analysis),*
  - Five why *(5 Why).*
- Identification of opportunities answers the following question: "what can be achieved?" and Identifies possible areas for improvement.
- The following methods are used to identify an opportunity:
  - checklist,
  - brainstorming.

2.4.2   Risk and Opportunity Assessment (Evaluation and Analysis)

- At the Biobank, the assessment of the risk and opportunity shall include:
  - determining the probability of occurrence;
  - determining the significance for the process in which it occurs;
  - determining the possibilities and methods of control.
- During the analysis, the Biobank shall:
  - determine the real level of risk/source of opportunity;
  - determine possible solutions resulting in the elimination of risk/describe the benefits of the opportunity;
  - identify the entities/areas to which the risk/opportunity relates.

8.5. ISO 20387:2018
4.1, 4.2, 6.1.2, 9.1.3 e), 9.3.2 e) ISO 9001:2015

The Most Common Practices

- As an example, the following three questions are used for risk assessment at the Biobank:
  - What can go wrong?
  - What is the probability that this will not work?
  - What could be the significance of the process (implications/effects)?
- Risk assessment is a comparison of the identified and considered risk with its criteria. In risk assessment, the strength of arguments collected in response to all three basic questions is taken into account.
- The effect of risk assessment is its qualitative estimation or a quantitative description of the risk size.
- Most often, Biobanks use the Risk List, which contains the following data: area, risk sequence number, first risk assessment date, risk title, risk level from the last evaluation, person responsible for risk evaluation, and person responsible for risk monitoring.
- To perform risk assessment, Biobanks use the Risk Assessment Form, which contains the following data: area covered by risk assessment, risk name, team appointed for risk assessment, risk assessment, and advised recommendations.
- As an example, the following three questions are used for the assessment of opportunities at the Biobank:
  - What can we achieve?
  - What is the probability that this will succeed?
  - What could be the significance of the process (possibilities for improvement)?
- Most often, Biobanks use the List of Opportunities, which contains the following data: area, risk sequence number, first risk assessment date, risk title, risk level from the last evaluation, person responsible for risk evaluation, and person responsible for risk monitoring.
- To assess opportunities, Biobanks use the Opportunities Assessment Form, which contains the following data: area covered by risk assessment, risk name, team appointed for risk assessment, risk assessment, and advised recommendations.

Frequently Asked Questions

- **How can risk analysis affect the functioning of the Biobank?**
  Risk analysis will show the advantages and disadvantages of the Biobank both in terms of the indicated technological process (e.g., the sample life cycle) and the entire Biobank system. When identifying our weaknesses by ourselves, we are able to counteract their negative effects in the future.

2.4.3 Reaction to Risk and Opportunity

- The Biobank shall identify and document improvement actions (reactions) in a given process.

The Most Common Practices
- One of the methods to qualify a risk response is to describe it, e.g., a reaction to a risk may result in:
  - reducing the risk (by designing and implementing control mechanisms);
  - minimizing the risk;
  - accepting the existing risk.

7.11.2.1, 8.5 ISO 20387:2018
6.1.1 a), 6.1.1 c), 6.1.2 ISO 9001:2015

2.4.4 Monitoring of Risk and Opportunity

- The Biobank shall establish a process of supervising the development of risk and opportunity control and verification of their level.
- The Biobank shall prepare relevant documentation of monitoring of risk and opportunity (reports).
- The Biobank should check and evaluate the effectiveness of the actions taken at defined intervals.

7.11.2.1, 8.5.2, 8.6.1 ISO 20387:2018
6.1.2 b2), 9.3.2 e) ISO 9001:2015

References

ISO 20387:2018 Biotechnology – Biobanking – General Requirements for Biobanking.
ISO 9001:2015 Quality management systems - Requirements.

Documentation and Records

3

Abstract

Documentation in the quality management system is a tool for controlling its operation. The use of the QMS documentation is advantageous to the Biobank mostly due to the unification of methods used by the employees in specific processes, thus reducing the occurrence of errors in both main and additional processes. The development of documentation makes it possible to organize the division of tasks and to establish the proper order of their performance, to define the scope of work for each employee, and to control the operation of processes and their optimization. The chapter *Documentation and Records* describes how to create the documentation and ensure its traceability, readability, accessibility, and confidentiality. It also refers to change control, backup, implementation, updates and withdrawals, distribution, storage and verification of documents covered by the QMS.

3.1   General Requirements

- The Biobank shall maintain internal documentation covering all aspects of its operations.
- The Biobank shall develop, implement, and apply a procedure to supervise the documentation included in the QMS.

  8.1.1, 8.2.1 a), ISO 20387:2018
  4.4.2, 7.5.1 b), 7.5.2, 7.5.3.1 ISO 9001:2015

The Most Common Practices
- The *Supervision over documentation* procedure specifies the traceability of documentation, readability of documentation, availability of documentation, confidentiality, supervision of changes in documentation, backup of QMS

documentation (creating a backup to restore/recover data after a possible loss or damage to some or all documentation) and elimination of backup copies of documentation.
- The QMS documentation includes regulations, procedures, instructions, templates for prints, labels, schedules, specifications, etc.

Frequently Asked Questions
- **What is the purpose of the documentation supervision procedure?**
  The purpose of the procedure is to ensure proper conduct in the field of oversight of the QMS documentation at the Biobank and to avoid the use of outdated or withdrawn versions of documents.

3.1.1 Traceability of Documentation

- The Biobank shall use a date format in all documents that complies with ISO 8601: 2004 (e.g., YYYY-MM-DD).
- The Biobank shall specify rules for the unique marking of documentation (applies to versions of currently applicable QMS documents and the withdrawn versions).

8.3.2, 8.4.2 ISO 20387:2018
7.5.2 a), 7.5.3.2 a), 8.5.1 a) ISO 9001:2015

The Most Common Practices
- *Exemplary labeling of document index:*
  System procedures are marked with the PS-XX symbol. The designation after the "PS" symbol corresponds to a specific procedure number. Derivatives from the print procedure have the following designation: a number originating from the procedure/symbol "S" from the system procedure/next printing number; (e.g., procedure PS-01. Supervision over documentation has an annex—print 1S-1 List of supervised documents). Technological Procedures (PT-XX), Equipment Procedures (PU-XX) and documents derived from them (printed matter, instructions, etc.) are created as specified above, so, to be specific,: PT-01 *Procedure for collecting biological material.* Annexes to the PT-01 procedure are, e.g., IN-PT-01-01 *Instructions for peripheral blood collection;* (in words: instruction no. 1 to technological procedure no. 1) or *1T-01 Report on collection of biological material;* (in words: technological printing no. 1 to technological procedure no. 1). One may also come across annexes labeled as follows: e.g., *Annex No. 1 Report on the collection of biological material for biological material Collection Procedure.* An annex labeled as such is much harder to find in electronic form; the documents will not fit properly according to the names. In addition, it is easier for the user to use short annex names, so using the main procedure number in annex names makes it easier to use QMS documentation on a daily basis.

- Each system document is accompanied by information about the date from which documents is valid and the date of withdrawal of the document and should also include the revision date or term or expiry date. Such information may be presented in tabular form: effective from . . .; withdrawn from . . . .

Frequently Asked Questions
- **If the Biobank has already introduced its system for labeling documentation, does it have to adapt it to the above-mentioned suggestions?**
  If your system for labeling documentation allows for the unambiguous identification of documents, the above requirement is met and there is no need to make any changes.
- **Can documents be distinguished based on, e.g., colors of the header (different color for system procedures, different for technological procedures)?**
  It is allowed for Technical and Technological Procedures and instructions derived from them to have a different design than others. Using colors is allowed. The division must be described in the Supervision of Documentation procedure.

3.1.2 Readability of Documentation

- The Biobank shall define rules for creating QMS documentation, define its appearance (graphic design, headline, footer) and the form ensuring the readability and user friendliness of QMS documentation.

  7.5.2 b), 7.5.2 c), 7.5.3.1 a), 7.5.3.2 b) ISO 9001:2015

The Most Common Practices
- There are several basic points specified in regulations and procedures, i.e.:
  - purpose and scope of procedure;
  - definitions and abbreviations;
  - annexes;
  - responsibilities;
  - implementation of procedure;
  - supporting documents.
- Instructions have several basic points, i.e.:
  - scope of instructions;
  - proceeding—performance of activities;
  - materials and equipment necessary to perform work (applies to Technology Instructions).
- Example of procedure PS-02 *Internal audits:*
  - **Objective:** Internal and external audits as well as documenting them.
  - **Scope:** Planning and conducting internal and external audits.
  - **Definitions and abbreviations:** Basic definitions are recommended, e.g. Record, Audit, Audit evidence, Audit Criteria, Auditor Team, Auditor's Competences, and QMS.

- **Annexes:** For example: Schedule of internal audits; Internal audit report; Auditor's competences; List of internal auditors.
- **Responsibilities:**
  Person responsible for updating this procedure:
    Person responsible for the QMS;
    Person responsible for approving this procedure:
    Biobank Manager—acceptance of the content;
    Person responsible for the QMS—acceptance of the content and formal approval;
    Persons responsible for using this procedure:
    Biobank employees and associates;
    Person responsible for verifying this procedure:
    Person responsible for the QMS.
- **Implementation of procedure**—here it should be described how to plan audits, how to prepare for an audit, how an internal audit is carried out, how reports are drawn up and audit results should be documented.
- **Supporting documents:** Examples of procedures: *Supervision over Quality Management System documentation*; *Identification, Reaction, and supervision of nonconformities*; *Continuous improvement process.*
- The purpose of the procedure is to determine the manner of conducting audits. Regulations, procedures, and instructions contain the following data: name and surname of the author of procedure and his/her signature, as well as the name and surname of the person approving the document and his/her signature. In addition, they should include the date from which the document is valid, the date of withdrawal of the document, version number, copy number, and name and designation of procedure; it should also include the revision date or term or expiry date.

- Prints contain the following data: date and signature of person preparing the document, date and signature of approvers, date from which the document is valid, and the date of withdrawal of the document.
- Documents cannot be handwritten; they cannot contain any deletions, corrections, or additions.
- The document footer clearly indicates with which document we are dealing.
  - **Example of a footer for regulations and procedures:**
    Copyright © 2017*, BIOBANK **, All rights reserved Page 1 of X ver. 1.00 YYYY-MM-DD****
  - **Example of a footer for prints and instructions:**
    Copyright © 2017*, BIOBANK **, All rights reserved Page 1 of X PS-02*** ver. 1.00 YYYY-MM-DD****

Key
* Year of implementation of the first version of document. When updating a version in the following year, it is obligatory to enter the year of creation of the first version and the last one. E.g., the first ver. 1.00 was created in 2006, ver. 1.01 was created in 2009, and version 1.02 was created in 2017. In this case, the

current version from 2017 should include the following: Copyright © 2006–2017.
** Biobank name (Name of the author and entity responsible for maintaining the QMS).
*** The number of documents which accompany a given document as "annexes." As a rule, we do not include procedures in the footer unless the procedure is an annex to a parent document which is established by regulations. However, when printing forms and instructions that constitute annexes to procedures, the number should be entered in the footer.
**** Date of implementation (validity) of the document.

3.1.3 Availability of Documentation for Biobank Employees

- The Biobank shall define and determine levels of access to documents for authorized persons (distribution of copies of documents).
- The person responsible for the QMS or an authorized employee is responsible for a correct distribution of documents, i.e., for distribution of only documents approved and valid at the Biobank.
- Current versions of documents shall be available for Biobank employees in electronic or paper form.
- Versions of withdrawn documents shall be properly cataloged (see Sect. 3.2.2 MBQM).
- The Biobank shall specify the way of sharing QMS documents with authorized persons.

8.2.6, 8.3.2 ISO 20387:2018
7.5.3.1 a), 7.5.3.1 b), 7.5.3.2 ISO 9001:2015

The Most Common Practices
- The use of the IT system at the Biobank facilitates documentation management. All changes, updates, approval of documents, or employee training are recorded in the system.
- It is allowed to provide disk space so that the current versions of documents are only available to employees in electronic form, in a format that prevents editing, on the access path—common space, e.g., I:/QMS (applies to all employees of the Biobank). Individual employees should be authorized to edit/view individual documents.

Frequently Asked Questions
- **How to distribute prints to make records during laboratory work?**
  The print original is available in a non-editable form on a disk accessible to all employees; if necessary, each employee can print it.
- **Is the print on which records are made during laboratory work required to have**

a table describing who created the print and who approved it?
No, only the originals of these prints include information regarding who created and approved the print.

3.1.4   Confidentiality

- The Biobank shall specify rules for maintaining professional secrecy regarding the internal documentation of the Biobank, taking into account the superior provisions of the unit/institution/organization in which it is located (including in accordance with the institution's security policy).

4.3 ISO 20387:2018
7.5.3.1 ISO 9001:2015

The Most Common Practices
- It is a good practice that every QMS document printed out of the system or disk is marked with information that it is an unattended printout/document, e.g., in the form of a watermark.

Frequently Asked Questions
- **Can I mark uncontrolled documents by hand?**
  Yes, e.g., after printing the document, it should be marked with a stamp with the inscription "unattended copy."
- **Can the university-wide information security procedure be considered sufficient so that the Biobank does not have to implement its procedures?**
  Of course, the Biobank may apply the entity's regulations and, when complying with it, does not have to implement its additional procedures.

3.1.5   Supervision of Changes in Documentation

- The Biobank shall specify rules:
  - for making changes in documentation and monitoring changes;
  - regarding assessment and approval of documents by authorized persons before their release.

8.2.1 b), 8.2.1 c) ISO 20387:2018
7.5.2 c), 7.5.3.2 c) ISO 9001:2015

The Most Common Practices
- It is recommended to use versioning of documents that significantly facilitates control over new versions of documents. It is recommended that minor changes (changes not affecting the change of the process) be included after a period, e.g., v.1.01, v.1.02, v.1.03, while large changes (e.g., technological changes affecting

the process) result in a change of the document version, e.g., from v.1.00 to v.2.00.
- It is a common practice to create records (register of changes in a document), as a result of which it is easy to find what has changed in a given document and why (introducing changes in processes, e.g., changing storage conditions of biological material, may be significant).

Frequently Asked Questions
- **Why may changes in QMS documents be introduced?**
  The introduction of changes in QMS documents may result, in particular, from organizational changes, changes in legal regulations and standards, from audits, management reviews, corrective actions, actions taken as a result of risk analysis, reviews of current documents, and changes in processes.

3.1.6 Backup/Safety Copies of Documentation

- To ensure the safety of documented information, the Biobank shall specify rules for making backup copies of the current and withdrawn QMS documentation.
- The process of making backup copies of and storing the QMS documentation shall be carried out in a way that ensures protection against damage.
- Regular tests confirming that it is possible to recover data from backup copies shall be performed. Tests shall be documented.

8.4.2, 8.4.3 ISO 20387:2018
7.5.3.1, 7.5.3.2 b), 7.5.3.2 d) ISO 9001:2015

The Most Common Practices
- Archive copies are stored in a location other than the location in which the original data are located.
- Backup and archive copies are made on portable high-durability media or on equipment that ensure adequate durability and safety of recording.
- Carriers containing backup copies must be protected against unauthorized access.
- Rules regarding the period of storage of backup copies of QMS documentation and their disposal should be laid down.
- Damaged backup media containing QMS documentation should be disposed of in a way that prevents recovery and reading of data from them.
- Elimination/destroying should consist in blurring data by recording other data on the same carrier or in a physical destruction of the medium.
- Some Biobanks have concluded a contract with external companies to perform services of destroying storage media. The most important thing is that companies offer safe and lawful disposal. The contract of entrustment, the purpose of which is to destroy documents containing personal data, should indicate organizational security, and provisions prohibiting the contractor from further entrustment. It should be noted that the contractor cannot entrust data to another entity. The

contract should also specify the manner of destroying documents in order to ensure that the entrusted data cannot be recovered. Any destruction of the storage media should be confirmed in writing.

3.2   Implementing Documentation

3.2.1   Responsibility for Dealing with Documentation

• The Biobank shall specify the responsibility for dealing with QMS documentation in the following areas:
  – preparation;
  – updating document content;
  – approval for use;
  – distribution;
  – use;
  – verification.

5.9 ISO 20387:2018
7.1.2, 7.5.3.2 ISO 9001:2015

The Most Common Practices
• The document may be prepared by any designated employee who has appropriate knowledge in a given field
• The document update is made by a person responsible for a given process
• Approval of documents:
  – an approving person responsible for the content of the document and for the functioning of the process;
  – an employee responsible for the QMS at the Biobank, who approves the document formally and implements the document to the QMS.

Frequently Asked Questions
• **If there is no designated liability in the current quality system, do I have to designate it for every document at the Biobank?**
  No, liability is only specified in the procedure that sets up instructions and forms.

3.2.2   Update and Withdrawal of Documentation

• The Biobank shall define rules for updating, making changes in and withdrawal of QMS documentation.
• The Biobank shall document the process of updating QMS documentation.
• The Biobank shall document the process of updating and withdrawing outdated versions of QMS documentation.

- The Biobank shall designate a person responsible for withdrawing QMS documentation.
- The implementation of changes to QMS documentation shall be associated with the obligation to provide training for all employees responsible for compliance with its content and/or its provisions.

8.2.1 b), 8.2.1 c), 8.3.2 b), 8.3.2 c), 8.4.2 ISO 20387:2018
7.5.2, 7.5.3.2 c), 7.5.3.2 d) ISO 9001:2015

The Most Common Practices
- It is a common practice to add an "Index sheet for changes to QMS documenta-tion," containing the following information:
  - number of previously revised document versions;
  - date of document withdrawal (in the YYYY/MM/DD format);
  - description of the change made.
- The person responsible for the QMS shall update system procedures, while documentation covering biobanking process procedures as well as equipment procedures or records is the duty of a person responsible for the implementation of individual processes at the Biobank.
- It is a common practice to annotate the original document with information about its withdrawal and to label the archive copy (e.g., annotation in the "QMS documentation index of changes" and on the cover page of the withdrawn document—it could be an annotation in the "QMS documentation index of changes").

Frequently Asked Questions
- **What to do if a new version of the document is issued but an employee is absent, e.g., on sick leave or vacation?**
  It is advised to provide training for the employee immediately after his/her return to work with the current date.

3.2.3 Training After Documentation Updates

- The Biobank shall define rules for documenting employee training in the field of new updated QMS documents.
- During the physically conducted training in new versions of documents, signatures of trained persons confirming that they have read the content of the document shall be collected.

6.1.1, 6.2.3.1 ISO 20387:2018
7.5.3.1 a) ISO 9001:2015

The Most Common Practices
- It is a common practice to use the annex to the system procedure *Supervision over QMS documentation*—print *List following QMS documentation training*, which should contain the following information:
  - document number;
  - version number;
  - number of pages in the document;
  - name of the document concerning which training is carried out.
    In addition, the following record should be included: *I declare that I have read the above document and undertake to apply it*; First and last name of the trained person, date and signature.
- It is a common practice to distribute QMS documents electronically for training in existing new documents. Documents should be saved in a non-editable format.

Frequently Asked Questions
- **How to train employees working outside the Biobank building, i.e. those who sample biological material?**
  It is allowed to conduct employee training by sending documents in non-editable electronic form. To use this mode of operation, you need to prepare a procedure describing this particular method of training (electronic).

3.2.4 Documentation Distribution

- The Biobank shall specify rules for documentation distribution:
  - the person responsible for the QMS or an authorized employee is responsible for the distribution of documents at the Biobank.
- The Biobank shall specify rules regarding the availability of documents, rules of printing and distribution.
  - documents must be distributed after their formal approval without undue delay;
  - it is forbidden to distribute documents before training the employee whom this process concerns (If an employee is trained in electronic form, the training can take place simultaneously with the distribution of documents).
- The Biobank shall define rules of supervision of distributed copies of documents.

8.1.2 b), 8.1.2 c), 8.2.1 c), 8.4.2 ISO 20387:2018
7.5.3.2 a) ISO 9001:2015

The Most Common Practices
- Copies of documents (unattended) are not stored.
- The distribution takes place in paper or electronic form (a document in a not-editable format)—it depends on the implemented form of managing QMS documentation.

- The employee is obliged to check whether the version of the document has not been changed before starting work.

Frequently Asked Questions
- **Is it possible to distribute internal Biobank procedures/instructions to stakeholders, e.g., collecting biological material?**
  Yes, in accordance with the accepted way of communication with a given entity, e.g., issuing an unattended paper copy, sending an electronic version of the document, as agreed with such entity.
- **How to inform stakeholders about changes in their current document?**
  Such information should be provided in accordance with the accepted communication rules.

3.3   Storage of QMS Documentation

- The *Supervision over QMS documentation procedure* at the Biobank shall specify rules regarding:
  - the manner and place of storing documents;
  - a list of documents that are currently in force and that have been withdrawn;
  - storage time:
    - (a) documents withdrawn from the QMS;
    - (b) medical documentation (if present at the Biobank);
    - (c) records made during processes (e.g., 1-T-01 *Report on the collection of biological material*).
- The Biobank shall ensure that QMS documentation is available whenever it needs to be used.
- The Biobank shall ensure the physical safety of stored paper versions and electronic documentation (original documents shall be kept in a safe place, e.g., in a locked cabinet).

   4.1.8, 8.3.2 f), 8.4 ISO 20387:2018
   7.5.3.2 b), 8.1 e) ISO 9001:2015

The Most Common Practices
- It is a common practice to register documents (e.g., in the form—*Register of documents and supervised, issued, and withdrawn entries*). The register records the date of validity of a given document version, as well as information about the currently applicable version, related documents/annexes, supporting documents and persons responsible for the implementation of a given process.
- The originals of the approved documents are in possession of the person responsible for the QMS and are made available to inspection entities by the person responsible for the QMS or a person designated/authorized by him/her.
- QMS documentation and medical documentation are stored in files or thematic files with a description of their contents.

- It is a good practice to place outdated documents in a binder marked "Withdrawn documents" and keep them in a different place than the current (present) documents.

3.3.1 Records

- The Biobank shall specify rules regarding storing records (in paper and/or electronic version).
- The Biobank shall specify the duration of storing records in accordance with applicable legal requirements (if applicable).
- The Biobank shall specify the manner of storing records (ensuring that they will not be destroyed/damaged/changed).
- The Biobank shall ensure the availability of records (availability in any situation whenever there is a need to use them).
- Actions requiring recording, circulation, and archiving shall be specified in individual procedures.

6.4.1.2, 6.4.1.6, 8.2.4, 8.4.1 ISO 20387:2018
8.1 e) ISO 9001:2015

Frequently Asked Questions
- **What should be done with records regarding the work of equipment that may become illegible?**
The best practice is to make a photocopy or scan of the document (e.g., a printout from an autoclave).

3.4   Verification of Documentation

- The Biobank shall determine how often QMS documentation will be verified to ensure that it is actually correct/up to date.
- The Biobank shall make records of documentation verification.
- Responsibility for the regular verification of QMS documentation rests with the person responsible for the QMS (see Sect. 3.2.1 MBQM).

8.2.1 b), 8.3.1, 8.3.2 b), 8.3.2 c) ISO 20387:2018
7.5.2 c), 7.5.3.1 a) ISO 9001:2015

The Most Common Practices
- It is a good practice to use a separate print (e.g., as an annex—*Verification of QMS document*), which can be established by the *Supervision over QMS documentation* procedure so as not to make changes to the document itself. The print should contain the following information: document number, document name, version number, and number of pages in the document.

3.4.1   Periodic and Incidental Verification of Documentation

- Periodic verification of documentation shall take place at specified intervals. The person verifying documentation should check whether the procedure described in the document is current and correct and whether the procedure described in the document complies with applicable law and accepted standards.
- Incidental verification of documentation shall take place after changing the law/guidelines/standards regarding Biobanks. The person who verifies the documentation should check whether the procedure described in the document is in accordance with currently applicable law and adopted standards.

8.3.2 ISO 20387:2018

The Most Common Practices
- An out-of-date content was noted during verification (a record having a key impact on the processes carried out). In this case, we do not extend the validity of the document and immediately update the document in terms of the above-mentioned record.
- During verification, an outdated record was noted (a record that does not have a significant impact on the processes carried out). In this case, we extend the validity of the document by, for example, a week and during this time we update the document.
- If ethical and legal requirements for informed consent change, check the Consent form and the procedure that describes the guidelines for creating consent and adapt their content to changes.

References

ISO 20387:2018 Biotechnology – Biobanking – General Requirements for Biobanking.
ISO 9001:2015 Quality management systems - Requirements.

Human Resources Management

4

Abstract

The Human Resources Management describes the human resource management policy in each well-organized entity. Regarding to this, the top management may employ competent staff with required qualifications as well as plan and organize training to build an effective and efficient QMS. The chapter *Human Resources Management* discusses in detail the requirements for identifying employees and hiring new staff, as well as all issues related to training the employees of the Biobank. As far as training is concerned, this chapter presents the issue of conducting initial and on-the-job training that enables employees to develop appropriate skills to participate in biobanking processes. It also describes development courses as well as the methods of verifying the knowledge of employees during the processing of biological material and related data. In the last part, the chapter *Human Resources Management* presents how the cooperation with a Biobank employee should be ended.

4.1 Human Resources Management Policy

- The legal entity, within which the Biobank is located, shall provide an appropriate number of qualified staff to carry out all biobanking processes.
- The Biobank shall define employee's qualifications for every position included in the Biobank's organizational chart.
- Employee's qualifications shall include education, training, or professional experience relevant to her or his duties at the Biobank.
- The Biobank shall create a job description that defines the scope of employee's duties at a given position with hazardous conditions. It shall also define the employee's role (function/position) and indicate responsibilities and authorities related to the Biobank.

© Wroclaw Medical University 2023
A. Matera-Witkiewicz et al., *Manual of Biobank Quality Management*,
https://doi.org/10.1007/978-3-031-12559-1_4

- The Biobank shall specify a replacement for each employee to ensure the continuity of all processes.
- The Biobank Manager shall clearly communicate the employee's scope of duties and obtain a written confirmation from the employee that she or he is aware of the resulting responsibility.

5.8 b), 5.9, 6.2.1.3, 6.2.1.4, 6.2.2 ISO 20387:2018
7.1.2, 7.1.4, 7.1.6, 7.2 ISO 9001:2015

The Most Common Practices
- Qualification requirements for an employee are usually prepared in the recruitment process by the manager of the legal or part of legal entity and determined for the specific position; e.g., for a biobanking specialist minimum requirements can be as follows: adequate education (biology, biotechnology, medical diagnostics), biobanking knowledge, and techniques related to the preparation of biological material.

Frequently Asked Questions
- **What are hazardous conditions?**
  Hazardous conditions are environmental factors that can be harmful to health. They can be divided into the following groups:
  - *biological factors: bacteria, viruses, fungi, other microbials, plants;*
  - *chemical agents: solutions of alcohols, acids, bases, solvents;*
  - *physical factors: vibrations, dust;*
  - *radiation: IR, UV.*
- **Is it possible to change the job description of an employee in the given position?**
  It should be done whenever a specific task is modified/introduced/deleted.

4.2   Biobank Employee Identification

- The Biobank shall possess and keep documentation allowing an employee to be identified.

4.1.7, 7.7.3 ISO 20387:2018

The Most Common Practices
- The documentation that identifies an employee may be prepared as a register and may include the first name(s) and surname(s) of an employee with the signature, signature's abbreviation, or internal employee identification.
- The Human Resources Department (or appropriate) stores documentation allowing an employee to be identified in accordance with applicable law.

Table 4.1 Identification of employee

Name, surname	Example of signature	Example of signature abbreviation	Internal identification code	Date of employment	Date of termination of employment
John Smith	*John Smith*	JS	*jsm*	2015-10-17	

- On the first day of work, a new employee may complete the form, where the signature and the signature's abbreviation are entered. The form can be as follows (Table 4.1):

Frequently Asked Questions
- **Is it possible to outsource the collection of employee identification documents?**
  It is possible if the Biobank is part of a legal entity where a proper department for dedicated duties is present. Each entity using internal rules regulates the organization of cooperation with the staff.

4.3   Admission of an Employee to the Biobank

4.3.1   Scope of Initial Training

- The Biobank Manager shall define rules for the admission of an employee to the Biobank structure.
- The legal entity, within which the Biobank is located, shall define the scope of training that a new employee shall undergo in order to provide the employee with the necessary information and inform him/her about the applicable rules at the Biobank.
- Biobank shall specify time requirements for the training of a new employee.
- In internal procedures, the Biobank shall clearly state that the employee will not be allowed to work without any training.

  6.2.1.5, 6.2.2., 6.2.3 ISO 20387:2018
  7.3 ISO 9001:2015

4.3.2   On-the-Job Training

- This type of training includes getting an employee acquainted with hazards present in his/her position, the possibilities of their prevention and protection.
- On-the-job training shall be performed according to the following stages:
  - the training participant gets acquainted with internal procedures regarding technical/technological processes appropriate to the position held;
  - the training participant observes Biobank employees during manual activities;

- the training participant performs manual activities under the supervision of a Biobank employee;
- the training participant takes a test verifying practical skills.

6.2.1.5, 6.2.2, 6.2.3 ISO 20387:2018
7.3 ISO 9001:2015

The Most Common Practices
- The Human Resources Department coordinates the admission of a new employee. Basic training before an employee is allowed to work includes:
  - initial training dedicated to safety procedures.
  - security and data security training for all Biobank employees, usually carried out by the Personal Data Controller (PDC);
  - QMS training for all Biobank employees and carried out by a person responsible for the QMS
  - IT training regarding the entire Biobank;
  - on-the-job training conducted by persons authorized to perform specific activities and to provide training for new employees. The Laboratory Manager defines on-the-job training specific to the laboratory.
- This basic training is conducted within the first month of work.

Frequently Asked Questions
- **Can I begin new employee training with the on-the-job training, which seems the most important?**
  A new employee should be informed about the principles of safety procedures always before job training begins.

4.4   Responsibility and Permissions of Key Personnel

- The Biobank management is responsible for aims/scopes and quality policy establishment and their communication as well as for providing resources; the Biobank management shall demonstrate leadership and commitment in relation to the QMS.
- The Biobank shall appoint a person or team responsible for the Quality Management System at the Biobank.
- Such person or team shall be responsible for the proper course of the process of implementation, maintenance, and improvement of the QMS.
- Such person or team shall be responsible for a periodic review of standard operating procedures, planning internal audits and reporting their results to the Institution's Top Management.
- The Top Management shall appoint a person or team responsible for an appropriate organization of technical/technological processes and also for QMS documentation verification and approval.

- The Biobank Manager is responsible for the effectiveness of the QMS and shall base management on a process approach.

4.1.7, 4.2.2, 5.2, 5.8, 5.9, 5.10, 7.4.5, 8.2.2., 8.3.2 a), 8.5.1, 8.6.1, 8.9.1 ISO 20387:2018
5.1.1, 5.1.2, 5.2.1, 5.3, 9.3.1 ISO 9001:2015

4.5   Biobank Personnel Training (Improvement Training)

- The Biobank shall establish and apply a policy regarding employee training. Training shall cover the following areas:
  - introducing a new method to the biobanking process;
  - introducing changes in technological processes;
  - renewing (refreshing) knowledge of the applied processes;
  - acquiring new knowledge in the areas of biobanking.
- Training in these areas can be carried out internally, i.e., by a Biobank employee (Biobank Manager, persons authorized by the Biobank Manager) or externally, i.e., by entities not related to the Biobank but having appropriate qualifications.
- The Biobank shall create a training schedule and update it.
- The training schedule should be prepared by a person responsible for the QMS and shall be accepted by the Biobank Manager.
- Any deviations from the training schedule should be recorded.

6.2.2., 6.2.3 ISO 20387:2018
7.1.6 ISO 9001:2015

The Most Common Practices
- The training schedule can be updated once a year or in any case if necessary.
- It is recommended that the Biobank Manager, together with the entire staff, estimates demand for training at the beginning of each year. After verification by the Biobank Manager, approval is given (Table 4.2).

Table 4.2 Example list of training courses

No.	Name and surname	Laboratory	Subject of training	Type of training	Quarter
1.	John Smith	DNA/RNA isolation	New DNA isolation technologies	External	Q1
			DNA isolation in columns according to the internal procedure	Internal	Q1
2.	Anne Smith	DNA/ RNA isolation	New DNA isolation technologies	External	Q1
			New RNA isolation technologies	External	Q2

Frequently Asked Questions
- **Who should train an employee after his/her absence and when he/she returns to work?**
  If an absent employee is employed as a member of the managerial staff, he/she may be trained by another manager participating in the training. If an absent employee manually performs a new technological procedure, he may be trained by another employee manually performing such new technological procedure.

4.6   Verification of Biobank Employees' Knowledge

- The Biobank should periodically verify the knowledge of employees manually performing the process to ensure that the process is performed in accordance with internal procedures.

6.2.2, 6.2.3.2 ISO 20387:2018

The Most Common Practices
- The knowledge of the employee who is performing a technological process can be verified during an internal audit; it can be also documented in audit reports or in the evaluation sheet.
- Once a year, a person who is responsible for the process is performing an exam concerning specific technological methods. The person observes an employee manually performing a technological procedure and at the same time verifies individual steps against the records of the internal procedure. After completion of the practical exam, he prepares a report with the results and presents it to the tested employee. The report is kept by the Biobank Manager.

Frequently Asked Questions
- **What should be done when an employee is seen to be making mistakes during the practical exam?**
  If the detected mistakes do not affect the quality and safety of biological material, the employee should be trained internally within the scope of recommendations of the internal procedure that applies to the process. The employee may still be allowed to work in the laboratory.

  If the detected mistakes affect the quality and safety of biological material, e.g., if they are connected with the traceability of biological material, the employee should stop working in the laboratory. Furthermore, training must be conducted, followed by another verification of the knowledge of such employee. Once the verification result is positive, the employee may be allowed to work in the laboratory again.

4.7 Training Documentation

- Participation in training should always be confirmed by a certificate or diploma if it was provided by an organizer. In the case of internal training, the attendance list is sufficient.
- Copies of certificates or diplomas confirming participation in Biobank employee training should be kept by the Biobank Manager or another authorized person.
- The Biobank shall verify the knowledge of internal training participants (Table 4.3) by using at least one of the following techniques:
    - written test;
    - oral questions;
    - practical exam.

6.2.3.1 ISO 20387:2018

The Most Common Practices

Table 4.3 Example of an internal document confirming internal training

Title of training			
Introduction to the Quality Assurance and Management System			
Place and date of training			
Wroclaw Medical University Biobank, December 12–15, 2017			
Scope of training			
Training in the provisions of internal procedures regarding the creation of procedures in the quality system and risk assessment.			
Name and surname of the person participating in the training	Signature	Date and time of training	
Anne Smith		12-12-2017	09:00-16:00
John Smith		12-12-2017	09:00-16:00
Name and surname of the person conducting the training	Signature	Date and time of training	Completion date and time
Olivia Smith			
Evaluation of training effectiveness			
The training was followed by a verification of the acquired knowledge and skills in the form of oral questions*. Training material has been mastered at the following level:			
Date and signature of the trainee	Date and signature of the person conducting the training		
* Verification method depends on the type of training, e.g. a written test, oral questions, a practical exam.			

4.8   Termination of Employment

- The Biobank shall specify and write down rules for delegating and transferring duties. The employee who assumes duties after the person with whom the employment contract is terminated shall have adequate qualifications (see Sect. 4.1 MBQM) and must undergo proper training (see Sect. 4.3 MBQM).
- The Biobank shall determine rules to remove permissions to access IT systems, data, and the entire Biobank infrastructure.

4.3.4., 6.1.1., 6.2.1.2 ISO 20387:2018
7.1.2 ISO 9001:2015

The Most Common Practices
- The employee with whom cooperation ends provides her/his superior with a written summary of:
  - the tasks conducted;
  - gives her/his superior or an authorized person all valuable objects that have been entrusted to her/him;
  - gives all keys and access cards.

  All access cards, user accounts in IT systems and room codes should be blocked.

References

ISO 20387:2018 Biotechnology – Biobanking – General Requirements for Biobanking.
ISO 9001:2015 Quality management systems - Requirements.

Ethical and Legal Aspects

5

Abstract

The fifth chapter is dedicated to ethical, legal, and societal aspects of biobanking that can be important from the point of view of quality. It offers recommendations on the principles of impartiality, confidentiality, and respect for autonomy. Detailed rules regarding informed consent, opinions of research ethics committees, the sharing of biological material and data, the commercial use of samples and data, ownership issues, returning results, and privacy protection were also described. Biobanking that complies with relevant ethical and legal standards may increase the quality and public trust in science and research.

5.1   Compliance with Ethical and Legal Requirements

- The Biobank shall comply with relevant regional, national, and international ethical principles and legal regulations for biological material and associated data. The Biobank shall have all necessary opinions, decisions, permits, and notifications required by binding law and should implement good practices according to non-binding legal guidelines and ethical recommendations.
- Biobank shall document its activities.
- The Biobank shall specify the retention period for records and related data.
- The Biobank shall respect human dignity and fundamental human rights and freedoms, particularly the right to privacy and respect for autonomy.
- The Biobank shall maintain impartiality in organization and management and analyze risks to impartiality.
- The Biobank shall be legal entity or be a part of an incorporated entity.
- The Biobank shall provide participants with access to full information about their rights connected with biobanking.
- The Biobank shall analyze and assess the risks for the privacy and autonomy of participants and pass such information to participants.

© Wroclaw Medical University 2023
A. Matera-Witkiewicz et al., *Manual of Biobank Quality Management*,
https://doi.org/10.1007/978-3-031-12559-1_5

- Documents confirming compliance with the provisions of national law, international guidelines, and ethical recommendations shall be cataloged and available to recipients/Biobank's stakeholders.

4.1.3, 4.1.4, 4.1.6. 4.2.1, 4.2.2, 4.2.3, 4.2.4 ISO 20387:2018

The Most Common Practices
- Placing legal bases for Biobank establishment and operation on the website (a copy or links to legal acts, legal regulations, etc.).
- Placing ethical bases for Biobank establishment and operation on the website. Depending on the country they can include: national or international ethical guidelines, recommendations, good practices and a copy of the Research Ethics Committee [REC]/Bioethics Committee [BC] approval.
- Placing a copy of the authorities' (Rector's/Unit Manager's) order in which the Biobank is included in the structure of the institution on the Biobank website.
- Placing a copy of Research Ethics Committee [REC]/Bioethics Committee [BC] approvals regarding currently conducted projects using samples, data, and other Biobanks' resources on the website. Providing information about the donor's rights on the Biobank website or in the form of an information leaflet.
- Information should be written in a simple and understandable language.
- Placing the most frequently asked questions along with answers to them (Q&A) on the Biobank website.
- Guidance on social responsibility: see ISO 26000:2010.

Frequently Asked Questions
- **Is a special legal act necessary for the Biobank to be established and operate?**
  No, Biobanks for research purposes may be established based on the principle of freedom of scientific research in accordance with appropriate national and international regulations and under the supervision of competent authorities. In different countries, there are different models of legal regulation of biobanking for research purposes: acts specific to a national Biobank, laws on Biobanks/biobanking, general laws on human biological material, general laws on scientific research, law regarding the collection and marketing of biological material and regulating laboratory performance or other medical law acts.

5.2   Ethical Advice and Approval

- The Biobank shall ask for advice or apply for an ethical approval for research activity to advisory boards and the REC/BC.
- The Biobank shall have an advisory board concerning: (1) scientific issues and (2) ethical, legal, and societal issues (ELSI).
- Biobanks, who do not possess theirs own Advisory Board shall ask for advice to other Advisory Board situated at National Node.

- All research projects on human biological material and/or associated data conducted at the Biobank shall be approved by an appropriate ethical committee (according to the national and international provisions and recommendations).
- The Biobank shall have relevant and up-to-date opinions/approvals of the REC/BC regarding research in which biological material and data from the Biobank are used.

5.3 ISO 20387:2018

The Most Common Practices

- National Biobanks, national nodes, or large Biobanks should establish their own scientific and ELSI advisory boards which should be different than any Research Ethics Committees/Bioethics Committees (REC/BC). Advisory boards should support the Biobank in current and planned scientific activities and policies, developing procedures and audits. Then REC/BC should be asked for advice on a specific research project.
- Smaller Biobanks or biorepositories may establish their own scientific and ELSI advisory boards or may apply for advice to scientific and ELSI advisory boards established at national Biobanks or national nodes.
- A network of REC/BC (name according to the country) is a way of exercising ethical, legal, and social control over the process of scientific research.
- An opinion of the REC/BC is required when:
  - a new project involving resources of the Biobank is planned;
  - changes to a research project are planned (the opinion should be obtained before the changes are implemented), e.g.:
    (a) changing the scope of consent;
    (b) modifying the type or amount of collected/stored biological material or data.
- The Biobank should apply for the opinion of the REC/BC in doubtful situations related to processing, using, and access to biological material and data, which raise concerns regarding participants'/donors'/patients' or recipients' rights, e.g.:
  - the scope of consent is different from planned research;
  - the clinical relevance of incidental findings is difficult to determine;
  - proprietary and ownership issues.

5.3 Impartiality

5.3.1 Impartiality Safeguarding

- Impartiality shall be safeguarded within the organization and management of the Biobank.
- The Biobank management shall be committed to impartiality and avoid internal and/or external pressure(s) that could compromise impartiality.

4.2.1, 4.2.2, 4.2.3 ISO 20387:2018

The Most Common Practices
- Impartiality at the Biobank can be safeguarded by:
  - transparent policy (see Sect. 5.3.3 MBQM);
  - organizational independence in the structure of hospital/university;
  - financial independence;
  - respecting participants' rights and freedoms;
  - legal capacity of the Biobank.

Frequently Asked Questions
- **What does impartiality mean?**
  *Making impartial decisions at the Biobank means giving adequate consideration
  to the interests of all concerned parties. Decisions should be based on objective,
  just, fair, and equal criteria and not on prejudices or favoring the benefits of one
  person/one group over another (other). Impartiality at the Biobank is an ethical
  standard and an accountability procedure.*

5.3.2 Risks to Impartiality

- Risk to the Biobank's impartiality shall be analyzed, identified, documented, and
  minimized or eliminated on an ongoing basis.

4.2.4, 4.2.5 ISO 20387:2018

The Most Common Practices
- Risks to impartiality should be analyzed from management, financial, research
  participant recruitment and communication perspectives.
- Analysis, identification, and documentation of risks for Biobank impartiality
  should be carried out particularly in relation with:
  - Biobank establishment;
  - organizational changes;
  - new project, particularly:
    (a) commercial research projects;
    (b) research involving vulnerable minorities;
    (c) cross-border sharing of samples and data.
  - prognosed/expected changes in legal, ethical, and societal circumstances:
    (a) new national or international legislation comes into force;
    (b) connecting with new registers;
    (c) natural changes in participants' rights, e.g.:
        establishing rules for conduct with material after participant's death;
            rules for conduct with material and data after reaching puberty by
            children and youth (in pediatric Biobanks);
  - liquidation of the Biobank.

Frequently Asked Question

- **What circumstances threaten the impartiality of the Biobank?**
  A relationship that threatens the impartiality of the Biobank can be based on ownership, governance, management, personnel, shared materials and associated data, finances, contracts, marketing (including branding), payment of a sales commission, or other inducement for the referral of new users, etc.

5.3.3 Transparency and Documentation

- The Biobank shall conduct a transparent policy.
- The Biobank shall define its mission and make it available (see Sect. 1.2 MBQM).
- The Biobank shall document in a comprehensible format all information relevant to Biobank activities, processes, and procedures and make it available for authorities and stakeholders (see Sect. 3.1 MBQM).
- The Biobank shall define the time period for retention of documented information and associated data relating to each biological material, after a complete distribution, disposal, or destruction of such biological material (see Sect. 3.3 MBQM).

4.1.3, 4.1.4, 4.1.6 ISO 20387:2018

The Most Common Practices

- The transparent policy should be conducted with regard to:
  - avoiding real conflict of interest;
  - disclosing conflict of interest (e.g., by the principal researcher);
  - transparent rules for access to samples;
  - common rules for access to and using samples and data;
  - policy of the non-discrimination of donors/researchers;
  - transparent relationships:
    - (a) external: relationships between the founder/owner and the Biobank;
    - (b) internal: between the functional units within the Biobank/

5.3.4 Responsible Supervision

- The Biobank shall specify the responsibility, authority, and relationship of personnel who manage, perform, validate, or verify work affecting biobanking output. The management body or a person who has overall responsibility for the Biobank should be defined and clearly indicated.
- The Biobank shall be responsible for activities performed in its facilities.
- The Biobank shall have a procedure to address liabilities arising from its activities.

5.1, 5.2, 5.4, 5.5, 5.8 ISO 20387:2018

The Most Common Practices

- Internal oversight procedures should include:
  - establishing a procedure for safeguarding biological material and associated data in collecting/procuring, acquiring, tagging, accessioning/logging, cataloging/classifying, examining, preparing, preserving, storing, managing data, destroying, packaging as well as safeguarding, distributing, and transporting;
  - periodic (at least once every 24 months) reports to supervisory bodies in the structure of the institution regarding performed and planned activities, information on access to and sharing biosamples and data, research projects in which the Biobank resources are used;
  - risk analysis and verification, e.g. risk related to the collection, use of, and access to biological material and data in the context of impartiality and confidentiality (see Chap. 2 MBQM);
  - introduction of the participant satisfaction questionnaire to survey opinions on the Biobank (e.g., approach to the donor, assistance in completing medical questionnaires and informed consent, quality of answers to questions asked by the donor regarding the project);
  - analysis of feedback information (remarks, conclusions, opinions, satisfaction surveys) from donors of biological material and institutions cooperating with the Biobank;
  - periodic verification of implemented supervision mechanisms may be carried out through audits (see Chap. 13 MBQM) and process control (see Chap. 11 MBQM);
  - it should define and periodically verify internal oversight procedures for its processes (depending on the type of Biobank).

5.3.5 Communication

- The Biobank shall specify the principles of communication with the participants and society during the implementation of the biobanking project, during research projects conducted on its resources and its results (see Sect. 1.4. MBQM).
- The principles of cooperation between the Biobank and the participant shall be transparent and available to the stakeholders.
- The parties involved in biobanking shall be defined and included in the communication plan.

5.10 ISO 20387:2018

5.4   Confidentiality and Privacy

- The processing of data is an integral part of the biobanking for research purposes. The Biobank shall protect the right to privacy and the confidential information (both obtained and created during the performance of biobanking) of participants, recipients, and users, particularly during storage and transmission of data.
- The Biobank shall comply with provisions and recommendation protecting confidentiality published by national and international data protection authorities and ethical committees.
- The Biobank shall ensure that the collected samples and associated data will not be shared with third parties for the purposes other than scientific (e.g., insurance companies or employers will not have access to data).
- Each Biobank operating for scientific purposes is required to demonstrate that the data it processes are related to the scope of its scientific activity, including future scientific research, compliance with legally binding documents, contractual agreements, the Biobank internal commitments, and ethical approvals.
- If there is no specific law related to the processing of personal data in Biobanks, the Biobank shall join, if possible, to code of conduct for processing personal data for research by Biobanks.
- When conducting surveys, interviews, and questionnaires, the Biobank shall provide privacy for the surveyed person, protection of personal data, comfort and security.
- The Biobank shall establish the procedure for data protection and confidentiality. Particular safeguards shall be applied when biological samples and associated data are transferred abroad (see Chap. 15 MBQM).
- In MTA/DTA there shall be a clause included referred to confidentiality and data (non)identification (see Sect. 14.3 MBQM).
- Procedures for the processing of personal data, using medical records, sharing biospecimens and data, the access to the Biobank database shall be established and well-known by the Biobank's employees.
- Every person (both inside and outside of the Biobank) having access to confidential data of the Biobank shall be bound to confidentiality. The Biobank should manage data in such a manner that they are fully protected against loss (accidental or hacking) during storage, sharing, and transfer of samples and data. Broad guidelines in this area are included in the ISO 27001 standard (see Chap. 15 MBQM).
- The Biobank should inform participants how their privacy and confidentiality are protected.
- The published results of research based on samples and data obtained from the Biobank cannot enable the identification of the participant/donor.
- The Biobank shall ensure the protection of privacy of participants at every stage of the biobanking process.
- Biosamples and associated biomedical data processed, stored, and transferred shall not enable the identification of the participant.

4.3.1, 4.3.2, 4.3.3, 4.3.4 ISO 20387:2018

The Most Common Practices

- It should be specified who is the data curator and who holds the position of the data processor in the Biobank.
- The Biobank should define and document the scope of competences required with respect to persons involved in the Biobank activities regarding confidentiality.
- The Biobank should periodically update data access authorizations.
- The Biobank should apply for the opinion of the REC/BC in doubtful situations related to processing, using, and access to biological material and data, which raise concerns and/or potential risk regarding confidentiality (e.g., the scope of consent different from planned research).
- Sensitive personal data and genetic data under Article 9 (2) (j) GDPR may be processed by Biobanks for scientific research purposes if such processing is legal, if it is necessary for the implementation of the scientific objectives set, if it does not violate the essence of the right to data protection and provides appropriate, specific measures to protect the fundamental rights and interests of the person, whom the data concerns. The processing of sensitive data by Biobanks should be done in accordance with rules stated in Article 5 of GDPR.
- Information for participants regarding confidentiality should include:
  - who the administrator of his/her data will be;
  - the method of data protection, particularly of sensitive data;
  - the possibility of using them in the future;
  - the period of storing samples and associated data (including personal data);
  - the conditions pertaining to their accessibility;
  - the possibility of withdrawing samples and erasing data from the database;
    - (a) the right may be limited, e.g., if the samples were anonymized and/or transferred to other entities in unidentifiable form;
  - the rules for making samples and data available to third parties, particularly to commercial entities and foreign entities;
  - proceeding in the event of the liquidation of the Biobank.
- Biosamples and associated biomedical data (e.g., regarding physical, physiological or behavioral, demographic, anthropometric characteristics and laboratory test results) should be separated from information enabling the identification of participants (e.g., name, personal reference number).
- The Biobank can protect the identity of participants by:
  - the anonymization or pseudonymization of samples and data which are processed, accessed, and transferred (assigning a unique code and/or removing information enabling personal identification from the samples);
  - the Biobank should document the anonymization or pseudonymization of biological material and data by:
    - (a) providing firewalls between information enabling the identification of the participant and the recipients (researcher, another Biobank);
    - (b) safe storage of samples and data;
    - (c) restricting access to samples and data.

Frequently Asked Questions
- **Is the implementation of the ISO 27001 standard required in the Biobank?**
  The Biobank should have security measures and data security procedure established and apply it in practice. ISO certification is not required.
- **Are specific security procedures applicable to all Biobanks?**
  No, it depends on the type of Biobank activity. If a Biobank is an entity conducting medical activity and the acquisition of biological material and data occurs in connection with health services and pertaining to medical records, the procedures should consider general provisions (e.g., GDPR) and specific area provisions (e.g., derived from medical law) contrary to the situation where a Biobank is not an entity conducting medical activity (e.g., in this situation only general GDPR provisions should apply). Every Biobank should perform the risk analysis.

5.5   Respect for Autonomy

5.5.1   General Requirements

- The Biobank shall respect the autonomy of every participant.

The Most Common Practices
- Respecting autonomy in biobanking is manifested by the rules pertaining to obtaining informed consent, respecting the right to refuse and withdraw the consent, the right to know and not to know, the policy regarding re-contacting, incidental findings, and giving clinically relevant information.

5.5.2   Informed Consent

- The Biobank shall collect material and/or data after obtaining informed consent and guarantee the right to withdraw the consent.
- The Biobank shall implement the procedure of obtaining informed consent for the participation in the biobanking.
- The consent shall cover each stages of the biobanking such as collection, storage, processing, use, sharing, and transfer of samples and/or associated data.
- Informed consent shall be obtained and documented before the beginning of material and data collection.
- The voluntary act of consent shall be preceded by providing participants with information on issues important to them in an understandable form.
- Participants shall have the opportunity to ask questions and discuss issues important to them with representative of the Biobank.
- The Biobank shall ensure that participation in the biobanking projects is completely voluntary.
- Informed consent shall be documented by the Biobank.

- The purpose for which the material and data are collected shall be specified as much as possible.
- The Biobank shall allow the donor to determine the scope of consent.
- The Biobank shall establish a policy of renewal/extension of the scope of previously given consent, if applicable.
- Biological material from clinical procedures (e.g., postoperative waste material) shall be used for research purposes if patient have given clear informed consent for the use of his/her material for research purposes. Consent for research purposes shall be separate from consent to clinical procedure.

The Most Common Practices
- The Biobank should provide participants with information about:
  - objectives of the biobanking for research purposes;
  - the type and method of material collection;
  - benefits, risks, or discomfort associated with the biobanking;
  - the manner of samples and data protection, re-use and destruction. The possibility of re-contacting;
  - the policy pertaining to incidental findings and giving clinically relevant information;
  - the possibility of accessing to samples and/or data by other entities;
  - the possibility of transferring samples and/or data to other entities, particularly to commercial entities and abroad;
  - planned commercial activity;
  - voluntary nature of participation;
  - the right to refuse donation or withdraw the consent for biobanking samples and/or associated data without giving a reason and without negative consequences.
- Information for participants should be accessible for every interested person.
- The Biobank can place the required information regarding the participant's rights and freedoms on a website and in printed leaflets prepared for participants.
- The potential participant/donor of biological material should not feel forced to give consent.
- The Biobank should provide the donor with enough time to consider whether to participate in the project and minimize the possibility of coercion or unauthorized influence on the decision.
- The most common form of documenting the consent is a written form.
- The Biobank should enable the donor to read the informed consent form and detailed information materials regarding the project before collecting biological material.
- The informed consent form should contain information about the participant, the place and date of consent.
- The participant should confirm with his/her signature that he/she has received and understood information.
- When the Biobank cannot specify the type and purpose of future research projects at the moment of collecting biological material, the consent for long-term storage

of material and data and the possibility of its re-use in many research projects should be obtained from the participants.

- The Biobank may use various types of informed consent, e.g. broad consent, dynamic consent, tiered consent (see Glossary).
- The Biobank should provide the participant with the opportunity to consent to the use of his/her biological material only for selected type of research or refuse that his/her material be used in some studies (e.g., the creation of human chimeras, embryo research, cloning).
- Secondary use of clinical material for biobanking and research purposes without clear prior consent for research can be acceptable only if:
  - samples and data are anonymized;
  - re-use was accepted through a REC/BC positive opinion;
  - it is not against the law.

Frequently Asked Questions
- **Is it possible to use biological material which we are not sure that the donor agreed to use in scientific research?**
  In this case the REC/BC should be asked to give approval to the use of such material after its anonymization according to specific country law.

5.5.3 Re-Contacting with Participants

- The Biobank shall establish a policy of re-contacting with the participant, particularly in the case of longitudinal studies with many follow-ups.

The Most Common Practices
- Information about the possibility of re-contacting should be provided at the time of obtaining consent.
- If the donor of biological material does not agree to be re-contacted, this information should be recorded in the documentation.

5.5.4 Right to Refuse Consent and Right to Withdraw Consent

- The Biobank shall provide the donor with the right to refuse consent to the collection and processing of samples and personal data without giving any reason.
- The Biobank shall provide participants with the possibility of withdrawing consent at any time without giving any reason.

The Most Common Practices
- The Biobank should identify the situations in which the donor's right to withdraw consent may be limited and inform about it before the beginning of the study.

- Information about the right to refuse, the right to withdraw consent (and possible situations when the withdrawal of consent will not be possible) should be made available in any promotional/ information material regarding the project related to the biobanking.
- The right to withdraw consent may be limited, for example, if samples were anonymized and/or transferred in an anonymous form to other entities.
- The right to withdraw consent may include
  - prohibition of future contacts with the participant, but permission for further storage and use of previously obtained samples and data in any form permitted by law;
  - prohibition of future contacts with the participant and anonymization of all samples and data;
  - complete withdrawal of consent, including destruction of samples and data.

5.5.5   The Right to Know and Not to Know the Results

- The Biobank shall respect the participant's right to know and not to know about the clinically significant results for his/her health and life.

The Most Common Practices
- The Biobank should establish rules for providing participants with feedback about the results which are clinically important for their health and life.
- The Biobank should establish the policy concerning incidental findings; information regarding the right not to know the results of research should be available in any promotional/information material concerning the project.
- Advisory Scientific Committee or REC/Bioethics Committee/Advisory Ethical Board can help in the assessment which information can be clinically important for participants' health and life.
- Unverified results obtained during scientific research should not be made available to participants.

5.5.6   Consent from Participants Who Are Not Capable to Give a Consent

- When a patient is unconscious the consent shall be expressed by the proxy or guardian of the participant or Research/Bioethics Committee.
- When a participant is not able to give a consent in writing (e.g., a paralyzed person), the consent shall be obtained in an oral form.
- In the case when the Biobank plans to collect material and data from people who are incapable of performing acts in law, the procedure of obtaining consent from the participant's legal representative (legal guardian) shall be implemented.
- The consent given by the legal representative shall be documented.

- The Biobank shall ensure the well-being and safety of a minor at every stage of the project implementation.

The Most Common Practices

- The Biobank should ensure that the dignity of persons incapable of performing acts in law (e.g., minors, people with mental retardation, dementia patients) is respected.
- It should be documented and confirmed by no less than two witnesses who are not related to the parties involved in the biobanking.
- The legal representative must have complete information about the biobanking processes and planned research to the highest extent possible in order to give consent on behalf of the participant.
- The legal representative should take into account an opinion regarding donation of biological material, expressed by the participant according to the degree of maturity and the ability to understand the situation.
- Collecting material and research projects in which material and data from people who are incapable of performing acts in law are used, must be approved by the REC/BC and are acceptable to conduct when a comparable study is not possible to be performed on the material collected from persons capable of performing acts in law. The research results should contribute to improving the health of the participant or other people in a comparable health situation.
- The Biobank should make reasonable efforts to obtain consent for further storage and use of biological material and data from a minor after he/she has reached the age of majority.

5.6 Sharing Material and Data

- The Biobank shall establish the rules for sharing and transferring of samples and data to other entities as well as the procedure for rejecting the application for the access to its resources (see Sect.14.2 MBQM).
- Material and data shall be shared in an anonymized form.
- Confidentiality shall be protected during material and data transfer.
- Biological material and associated data (without the possibility of identifying the donor) shall be transferred to another country if it is in line with the consent of the participant or—in the absence of a documented consent, after receiving an approval from REC/BC and in compliance with national law.
- Material and data shall be shared and transferred when an appropriate level of protection of fundamental rights and freedoms is ensured by the law of that country or legally binding instruments and agreements for protecting rights and freedoms were implemented by parties involved in the transfer.
- The Biobank shall only release information regarding biological material and associated data according to the relevant agreements and approvals (see Chap. 14 MBQM).

- When the Biobank is required by law to release confidential information, the participant/provider/donor shall be notified of the information provided, unless it is prohibited by law.
- The Biobank shall respect the rights and freedoms of the participant, particularly confidentiality and autonomy, in cooperation with service subcontractors during collection, transport, distribution, research, storage, processing, and sharing of biological material and data.
- The Biobank shall assess the subcontractor's ability to provide appropriate level of protection of the rights of the participant. Such an assessment shall be made prior to taking any orders.
- The Biobank shall define the rules of cooperation with subcontractors and describe them in detail in the form of a written contract/agreement.
- The Biobank shall periodically verify the concluded contracts and modify their content when necessary.
- The Biobank shall store all documentation related to the concluded contracts.
- The Biobank shall determine the procedure for the transfer of samples and data collections in the event of termination of the activity related to the biobanking.
- Biological material and associated data received or acquired by the Biobank shall be segregated until legal, ethical, documentary, and quality compliance of biological material and associated data has been assessed and managed.
- The Biobank shall document all processes, where the responsibility are determined, even if it is not responsible for collection or sampling this shall be also documented (see Chap. 3 MBQM).

4.3.3, 7.3.2.4, 7.3.2.6 ISO 20387:2018

The Most Common Practices
- It is recommended to enclose to the available samples a statement that the Biobank holds patients' consents and about the scope of consent expressed by the participant.
- The transfer of biological material and anonymized data requires signing of a contract defining the responsibility of each party involved in the transfer (with special regard to security and transport).
- The Biobank that provides samples outside the country is responsible for making sure that the standards of protection in a given country are at a level not lower than national requirements.
- Data security arrangements should be documented and validated by the parties involved in the transfer.
- In order to contact with participants/donors/providers the Biobank should:
  - obtain the consent for re-contacting;
  - obtain the data necessary to establish a contact (email, address, telephone number).
- The contract/agreement should specify the mutual obligations of the parties as well as the methods of communication.

- In the case of key suppliers (e.g., outsourcing transport of biological material) the contract should specify: scope of work, proceedings in case of changes/exemptions, scope of possible supplier control/audit, liability, duration of contract, terms of its termination, quality requirements for all services covered by the contract/agreement.
- The following should be documented:
  - the power of attorney;
  - annexes to contracts;
  - attachments to contracts.
- Before making a decision regarding liquidation of the biobanking facility, a biological material transfer contract should be concluded with the institution that will take over the samples and data.
- In the context of the distribution and disposal of biosamples and data the following are required:
  - compliance documentation with applicable regulatory and ethical requirements;
  - verification documentation of biological material and of associated data;
  - documented agreement(s) and legally binding document(s).
- The Biobank should have a documented process of accepting or rejecting (in the case of noncompliance) of biological material collections and data sets from other Biobanks.
- The Biobank should make sure that the scope of consent from participants/donors pertaining to the acquired material and data allows for the transfer of samples and data.
- In the absence of such a consent the Biobank may apply to the REC/BC for opinion.
- Persons responsible for the biobanking in the acquiring institution should check the method of collecting material, their storage, processing, and transfer, as well as informed consents and relevant ethical approvals.
- The Biobank—the recipient of the collection, should obtain the copies of available documents certifying the techniques/research methods and the method of handling the sample during the biobanking process, as well as information on restrictions concerning its use.
- The acquired collection should be verified in terms of the quality of material on a representative sample set.
- The acquired collections, due to unknown history of samples, should constitute a separate set of data.
- Before acquisition, it is recommended to conduct an audit in the transferring center in order to verify the fulfillment of the requirements related to the biobanking.
- Checking the quality of acquired biological material by setting reference values for a given material and referring several samples of qualitatively important parameters against them.

5.7   Proprietary Rights and Commercialization

- The Biobank shall protect the proprietary rights of participants and stakeholders.
- The Biobank shall inform participants about a plan to conduct commercial activities based on biological material and data collected and participants shall consent to such activity.
- In the situation where, commercial entities are provided with material and/or data collected by a public or other non-profit entity, such information shall be made available to the participant before signing the informed consent form.

4.3.1 ISO 20387:2018

The Most Common Practices
- Proprietary rights should be addressed in the informed consent.
- The Biobank should protect users' rights, particularly intellectual property rights.
- In the event of unclear situations obtaining the opinion of REC/BC is recommended.
- In the event of unclear situations obtaining the opinion of REC/BC is recommended.

Frequently Asked Questions
- **Is it permissible to charge a fee for biological material provided?**
  Biobank has the right to collect fees related to the reimbursement of costs of collecting, processing, and storing samples of biological material. Each biobanking facility should determine the costs of obtaining samples of biological material depending on the type of collections.

References

ISO 20387:2018 Biotechnology – Biobanking – General Requirements for Biobanking.
ISO 9001:2015 Quality management systems – Requirements.
ISO 26000:2010 Guidance on social responsibility General Data Protection Regulation (GDPR)

Supplies, Materials Management

6

Abstract

Biobanking is a complex process that requires appropriate resources in the form of disposable materials and reagents, which have a direct impact on the biobanking output. To provide high-quality and repeatable biobanking output, the Biobank must ensure that such materials, mainly used in the processing of biological material, are properly supervised. This chapter *Supplies, Materials Management* discusses the requirements relating to the rules on the receipt of materials, their qualification, acceptance, and handling, as well as the requirements relating to the cooperation with suppliers, including their qualification, to ensure that the continuity of supplies and products provided by third parties does not adversely affect the activities of the Biobank. This chapter suggests the most common practices and examples of ready-to-use forms that are useful for documenting these activities.

6.1 Materials Receipt Rules

- The Biobank shall define and establish a policy regarding the management of materials that are used in the technological process of the Biobank.
- This procedure shall specify how the Biobank's materials requirements are communicated to the supplier.
- Materials management policy shall include the rules pertaining to the purchase, receipt, qualification, and acceptance as well as the documentation of these activities. It shall also specify the handling of rejected materials.

6.4.1.1, 6.4.1.5 ISO 20387:2018
8.4.2 ISO 9001:2015

6.2   Reception, Qualification, and Acceptance of Materials

- The Biobank shall determine the rules pertaining to the reception, qualification, and acceptance of materials.
- Reception, qualification, and acceptance of materials shall be documented (Table 6.1).
- Reception, qualification, and acceptance of materials shall be carried out by trained Biobank personnel.
- The following activities shall be carried out regarding reception:
  - verification of the order compliance with the delivered material;
  - visual verification of the packaging in which the material was transported and of the immediate packaging in which the material is contained.
- The following activities shall be carried out regarding qualification:
  - checking production date and expiry date;
  - checking the presence of the required accompanying documents;
  - checking transport conditions and further storage conditions.
- If the above activities are completed successfully, the material shall be qualified and accepted for use.
- If the above activities bring the unsuccessful results, the material shall not be qualified and accepted for use. It shall then be quarantined until clarified. After a positive explanation, the material is qualified for use. In the absence of explanation, the material shall be rejected.
- Qualification of material for use, transfer to quarantine or rejection shall always be documented with a description of the transfer to quarantine or rejection.

6.4.1.1 d), 6.4.1.5 ISO 20387:2018
8.1, 8.4.1, 8.4.2 ISO 9001:2015

The Most Common Practices
- Quality criteria are set for materials that are critical in the technological process, e.g., for cryotubes these are: sterility, tightness.
- In an emergency (such as no cryo tubes qualified and accepted for use), when the material is in quarantine, it may be accepted for use after being released by the Biobank manager. Release should be confirmed in writing.
- The status of the material during qualification is determined using colored labels, e.g. a green label means material accepted for use, a yellow label means material in quarantine, and a red label means material rejected.

Table 6.1 Example of reception, qualification, and material acceptance protocol

Name of the material or reagent:
Manufacturer:
Date of production / validity:
Series number:
Quantity:
Yes / No / Not applicable
❑ ❑ ❑ The number of the quality certificate according to the series number on the packaging
❑ ❑ ❑ The material / reagent according to the manufacturer's specification
❑ ❑ ❑ The material / reagent according to order
❑ ❑ ❑ Positive result of preliminary inspections
❑ ❑ ❑ The material / reagent provided with information leaflet
❑ ❑ ❑ The material / reagent transported as required

A.   Left in quarantine on:
 	
        Reason:
 	
B.   Qualified for use on:
 	
        Justification of dismissal:
 	
C.   Accepted for storage on:
 	
D.   Rejected on:
 	
        Reason:

Name and surname, date, signature:

Frequently Asked Questions

- **What is the verification of order compliance with the delivered material?**
  This verification consists in checking the name of the material supplied, the manufacturer, and the quantity ordered.
- **What can be the accompanying documents that should be verified when qualifying the material?**
  These can be any specifications (e.g., for the supply of antibiotics), safety data sheets, certificates (e.g., quality, analysis, or sterility certificates).
- **What is the quality certificate provided with the material?**
  It is a document confirming compliance with specific standards (e.g., ISO) or quality criteria (e.g., degree of purity) for a given batch of material.

- **What is the placement of material in quarantine?**
  It consists in separating the material and its clear marking in order to prevent its use in the technological process. Such material can be labeled with the word "Quarantine" and placed on a separate shelf intended for this purpose.
- **Does the reception, qualification, and acceptance report have to be a separate print?**
  No. Information can be noted, e.g., on the back of a copy of the order. It should be remembered that the entries contained therein allow reading the information required in point 6.2.

6.3   Qualification of Suppliers

- The Biobank shall define the rules for qualification (criteria) of suppliers and the frequency of this activity.
- A supplier's qualification shall be documented and stored (Table 6.2).
- A supplier's qualification shall serve to confirm the supplier's competence.
- The criteria shall clearly define the conditions for a positive qualification of the supplier for further cooperation with the Biobank.

Table 6.2 An example of a supplier assessment protocol

Supplier's contact details	"DNA/RNA" John Smith Wroclaw ul. Borowska 211, +48 1234567	
Name and surname of the evaluator	Anne Smith	
Position of the evaluator	Manager of DNA/RNA isolation laboratory	
Date of assessment	2017/12/22	
The assessment concerns:	Choice of a supplier ❏	Annual qualification of a supplier ❏
Evaluation criteria	The minimum amount of points: 9	
Punctuality (max 4 points)	2 points (delivery exceeded by 4 days than the time of delivery specified in an offer; delay acceptable)	
Completeness of delivery (max 4 points)	3 points	
Certificates/accreditations (max 4 points)	4 points ISO 9001:2015—required, ISO 20387:2018 accreditation—not required	
Sum of points	9	
Punctation	1. Does not meet the requirements 2. Does not meet all requirements, but is acceptable 3. Meets the requirements 4. Meets the requirements to a greater extent than expected	
Result of the assessment	Positive ❏	Negative ❏

- In the event of a negative qualification of a supplier, an authorized Biobank employee shall clearly communicate to employees dealing with orders about the disqualification of that supplier.
- The Biobank shall create and update a database of suppliers.
- The Biobank shall identify and assess the risk associated with materials provided by external entities. If necessary, the Biobank shall take measures to avoid any adverse effects that may result from the supply of material that does not meet the Biobank's requirements (see Sect. 2.4 MBQM).

6.4.1.2, 6.4.1.4 ISO 20387:2018
4.2, 8.4.1, 8.4.2, 9.1.3 f) ISO 9001:2015

The Most Common Practices
- It is a common practice that the supplier's base is updated once a year. The database can be kept in a paper or electronic form.
- New suppliers are added to the list on a regular basis.
- A new supplier can be qualified after the first delivery of materials.
- It is recommended to perform the qualifications of suppliers every year and summarize their all previous deliveries.
- An appropriate number of points should be assigned to each supplier evaluation criterion.
- A minimum number of points should be set, which clearly qualifies a given supplier for further cooperation with the Biobank.

Frequently Asked Questions
- **What can be the criteria for the supplier assessment?**
  The criteria are chosen by the Biobank itself and they can be: timeliness, completeness of delivery, price, certificates/accreditations/permits held, communication skills (including a response to complaints).
- **What happens if the supplier is negatively assessed by the Biobank, while ordering material from this supplier is obligatory and results from the Public Procurement Law?**
  In this case, you can contact the department responsible for public procurement in the unit where the Biobank operates and jointly influence the suppliers to improve the indicators that require it.

6.4   Handling of Materials Stored in Storage Space

- The Biobank shall determine and document storage conditions for materials. These conditions shall be in accordance with the recommendations of the material manufacturers.
- The Biobank shall set storage ranges, such as temperature and humidity (if required) as well as light protection.

- During storage of materials in the warehouse, the Biobank shall determine the frequency of verification of the above-mentioned parameters and document this verification. Operation shall be carried out to ensure stable storage conditions.

6.3.2, 6.3.4, 6.3.5, 6.4.1.4 ISO 20387:2018
7.1.4 c), 8.5.4 ISO 9001:2015

The Most Common Practices
- The designated person writes down the storage conditions for all materials, then divides the materials into groups with similar storage conditions and sets a common range of criteria.
- Storage parameters are monitored before the working day commences, in order to reduce the use of materials whose inadequate quality associated with incorrect storage parameters will contribute to the failure of the technological process.
- The material released for use from the warehouse should have the shortest expiry date available on stock—in accordance with the FIFO (First in, First out) principle.

Frequently Asked Questions
- **What storage conditions shall be considered before placing the material in the warehouse?**
  The following conditions shall be taken into account: temperature, humidity, the need to place the material in a dedicated safety cabinet (chemical cabinet), the need for separation from other groups of materials, e.g. acids and bases separately.
- **Where can I find information on the required storage conditions?**
  Such information can be found in leaflets, directly on the packaging and in the safety data sheets, which describe the safety rules in detail.

6.5   Handling of Materials Used in Daily Work

- The Biobank shall record the current use of materials in the technological process (see Sect. 8.1 MBQM).
- The Biobank shall create a list of materials taking into account the exact name, characteristics (e.g., capacity) and designate substitutes for individual materials, if possible.
- Reagents prepared internally at the Biobank shall be prepared in accordance with documented and validated methods (see Chap. 10 MBQM).

7.5.1 c) ISO 20387:2018
8.5.2 ISO 9001:2015

Table 6.3 An example list of materials

List of materials used in the process		
Name of the process: Addition of a cryoprotectant, DMSO		
No. of samples: AA12345678, AB12345678, AC12345678		
Name	Lot number	Expiration date
DMSO, manuf. XYZ	1234567	2020-06-15
1ml tips, manuf. XZY	8910111	2020-12-31
Cryo-vial CryoProbe, manuf. ZXY	9876543	2021-10-31
Name and surname, date, signature:		

The Most Common Practices

- After completing manual operations in the technological process, the employee creates a list of disposable materials and reagents used in the processing of a biological material, and confirms their use with a date and signature (Table 6.3). This document is stored.

Frequently Asked Questions

- **The Biobank received information from the manufacturer about a defect in disposable material, e.g., a cryotube with LOT number: XYZ123. How to identify if a given material was used in a technological process?**

  The Biobank should maintain and store a list of materials used in the technological process, including LOT numbers. After receiving information from the manufacturer about the defect of consumable material with a given LOT number, the Biobank identifies when and in which technological process the material was used. If it has not been used or accepted to use, the Biobank utilizes and/or exchanges the indicated batch. If the material was used in the technological process, then the manager of the Biobank or a person appointed by him, together with a person practically performing the technological process, assess the impact of the defect on the quality and safety of biological material.

- **On the basis of which validated methods shall the Biobank prepare the reagents internally?**

  The preparation instructions provided by the manufacturer should be considered as validated methods and may also be a method published in the professional literature. The Biobank may also independently validate the procedure for the preparation of the reagent, provided that this validation results in proving the stability of the prepared reagent and the date of stability. All methods of reagent preparation, regardless of their origin, shall be documented.

References

ISO 20387:2018 Biotechnology – Biobanking – General Requirements for Biobanking.
ISO 9001:2015 Quality management systems - Requirements.

Equipment

7

Abstract

Biobanks use a wide range of devices, from small equipment, such as pipettes, through freezers, to advanced and expensive apparatus, such as cytometers or sequencers. If properly planned and implemented, the equipment management process guarantees that the Biobank will be equipped with devices it needs to carry out its activities. Appropriate surveillance of the equipment makes it possible to ensure that the devices will function in accordance with the specifications, they will be safe for operation, and their service lives will be extended. In this chapter, we described the requirements for the acceptance of equipment into the Biobank (installation, operation, and operational qualification) based on the pre-defined criteria, important issues relating to its daily operation, including maintenance, records, documentation, continuous supervision, and work with devices that do not meet the requirements.

- The Biobank's equipment shall be adapted to the purpose of its functioning.
- The Biobank shall establish and maintain a procedure pertaining to the acceptance of equipment and their qualifications (IQ, OQ, PQ) before use (prior use), rules for their day-to-day usage and recording the above actions.
- The basis for the definition of rules connected with the Biobank's equipment shall be:
  - definition of the equipment user requirements specification prior to its purchase (required in the case of a critical equipment);
  - qualification of every equipment according to its purpose;
  - establish unique identification of equipment;
  - preparation of a list/register of equipment in the Biobank's equipment inventory with clear identification of critical equipment;
  - preparation of internal equipment specification.

© Wroclaw Medical University 2023
A. Matera-Witkiewicz et al., *Manual of Biobank Quality Management*,
https://doi.org/10.1007/978-3-031-12559-1_7

6.5 ISO 20387:2018
7.1.3, 7.1.5.1, 8.5.1 d) ISO 9001:2015

The Most Common Practices

- In the specification of the equipment which the Biobank intends to buy, describe selected parameters, e.g. in the case of a centrifuge—minimum speed, centrifuging temperature, and maximum possible amount of centrifuged material at the same time.
- The Biobank's equipment is assigned with its unique internal number. The Biobank may use a number assigned by the Administrative Department (the so-called inventory number), which uses the number to identify the equipment as a fixed asset. Also, the Biobank may use a number assigned by the manufacturer.
- Qualification of the equipment is described in points 7.1.2.1.–7.1.2.3 MBQM.

Frequently Asked Questions

- **The Biobank possesses the equipment approved without defined rules of acceptance. How to proceed with such equipment?**
  The Biobank should implement a procedure concerning the principles of daily use and the performance of the above-mentioned activities for the equipment used so far. For newly purchased equipment, the Biobank should implement and apply an acceptance and qualification procedure (IQ, OQ, PO) prior to commissioning. It is recommended to include the requirement to provide documentation confirming the qualification of the device (i.e., IQ, OQ) in the order for the device.
- **Should the Biobank qualify the equipment purchased before implementing QMS?**
  The Biobank should implement and apply the qualification procedure (IQ, OQ, PQ) for devices that were purchased before implementing QMS. This particularly applies to critical devices used in the technological process.

7.1   Acceptance of the Equipment

- The Biobank shall perform qualification according to adopted criteria (see Sects. 7.1.2.1–7.1.2.3 MBQM) for every critical equipment before prior use.
- After acceptance of the new equipment, following a repair or a software update, a renewed qualification shall be performed, based on the adopted rules and the manufacturer's requirements.
- Acceptance of the equipment is equivalent with the introduction of the equipment to the list/register of the equipment contained in the Biobank's inventory.
- An update of the said list/register is required to be performed at specified intervals and following each purchase of the new equipment.
- The Biobank Manager is responsible for the acceptance of an updated list/register of the equipment or a person indicated by the Biobank Manager.
- The Biobank is required to prepare internal user description of equipment (internal instruction) for the critical equipment, i.e. a document covering the following issues:

- name and type of the equipment and software (if applicable);
- name of the manufacturer and equipment identification data (e.g., serial No.);
- characteristics and intended use of the equipment (short description);
- the most important directions for safe usage (i.e., the manufacturer's recommendations);
- description of the process of cleaning and decontamination of the equipment;
- recommendations for maintenance and servicing;
- recommendations for qualification/requalification (if required);
- indication of a substitute equipment in the case of failure;
- recording of damages, irregularities in operation and repairs;
- for each type of equipment, safety information should be provided including associated risks in the operation of the equipment;
- description of working with the device.

6.5 ISO 20387:2018
7.1.3 b), 7.1.5.2, 8.4.3 b2), 8.5.1 d) ISO 9001:2015

The Most Common Practices
- Criteria for the acceptance of the equipment operation should be defined (e.g., operational parameters, e.g. temperature, humidity, etc.).
- Acceptance criteria may be indicated in the instructions or specifications provided by the manufacturer of the device. In the case of submitting reports confirming the qualifications of IQ/OQ, the Biobank may adopt equivalent acceptance criteria.
- After obtainment of positive qualification results, the equipment should be included to the laboratory equipment inventory.
- After obtainment of negative qualification results, the Biobank implements corrective actions.
- The internal instruction of device describes the rules for working with the equipment, i.e. cleaning, maintenance and servicing, qualification and requalification.

Frequently Asked Questions
- **What type of a document the internal user description of the equipment is?**
  Internal user description of the equipment is a document (an internal instruction) that contains all information pertaining to safe usage of the equipment, preparation of the equipment prior to the commencement of work, its cleaning and conduct in the event of a failure.
- **Where can I obtain input data for the preparation of internal instruction?**
  Input data for the preparation of internal instruction are contained in the instruction and all recommendations provided by the manufacturer of the equipment. Input data also comprise experience and knowledge of employees who worked with a given type of equipment.
- **Is the instruction provided by the equipment manufacturer sufficient for working with it?**
  No, because the manufacturer does not define the specifics of a process in the Biobank, i.e. does not define the acceptance criteria for the equipment operation.

7.1.1 Device Acceptance Criteria

- The Biobank, prior to the purchase of the critical equipment, should define its specification (see this chapter MBQM). The specification should describe characteristic features of the equipment, without which there is no possibility to perform process. The following should be indicated:
  - the necessary operating parameters;
  - the values of acceptable limits of defined parameters.

6.5.8 f) ISO 20387:2018
7.1.5.1 ISO 9001:2015

The Most Common Practices
- Prior to the purchase of the new equipment, a market evaluation should be performed in order to compare prices, terms and duration of warranty, critical features of the device.
- E.g., within a specification for a centrifuge, that is the critical equipment in a given process, the necessary parameters of the centrifuge's operation are: speed and temperature. Acceptance limits include:
  - minimum speed: 100g;
  - maximum speed: 3000g;
  - temperature: +4 deg. C.
- User requirements for a device are equivalent to a request for quotation submitted to potential suppliers.
- A device is qualified according to the criteria documented in the user's requirements. A device will receive a positive status (ready to use) after the fulfillment of the user's requirements.

Frequently Asked Questions
- **Should the Biobank define all criteria for the equipment acceptance at once, prior to its purchase?**
  Yes; the defined criteria should be justified in the implemented process; e.g. the Biobank requires the manufacturer of a centrifuge to deliver the equipment allowing for adjustment of time and braking speed after the finished centrifugation. In the case of material centrifugation in a density gradient, the above-mentioned condition is indispensable in obtaining the right phase division.

7.1.2 Acceptance or Rejection of the Equipment

- The Biobank is required to have documentation of IQ, OQ, and PQ, at least for critical devices.
- It is required that every qualification performed (IQ, OQ, PQ—see Sect. 7.1.2.1–7.1.2.3 MBQM) to be documented, verified, and approved by the Biobank's Manager or a person indicated by the Biobank Manager.

- It is required that every deviation from the adopted acceptance criteria and corrective actions shall be documented and approved by the Biobank Manager.

6.5.6, 6.5.8 f) ISO 20387:2018
7.1.5.1 ISO 9001:2015

The Most Common Practices
- IQ, OQ documentation is supplied by the manufacturer with the device.
- An example of a negative result qualification:
  - laboratory centrifuge:
    - (a) the diagnosed cause is a damaged centrifuge flap (the event was documented on the device acceptance protocol—IQ);
    - (b) you cannot go to the next step: OQ;
    - (c) contact the service center in order to replace the device or the flap with a new one (corrective action). The Biobank Manager accepted the deviation in ongoing IQ (damaged flap) and corrective action (contact with service and replacement of the damper or device);
    - (d) after replacing the damaged flap, IQ is carried out again, after receiving a positive IQ result, we start OQ and PQ.
  - applies to the device equipped with a computer and an external screen:
    - (a) the diagnosed reason is the lack of image (dark screen) on the device's display after starting it;
    - (b) you cannot go to the next step: OQ;
    - (c) contact the device service center in order to repair the problem or provide a new device. The Biobank Manager accepted the deviation in the course of IQ (display not working) and corrective action (repair or replacement of the device);
    - (d) after the repair of the device, IQ is carried out again, after receiving a positive IQ result, we start OQ and PQ.
- Installation Qualification performed with a negative result—the reason is the lack of power cable for the centrifuge (the lack of cable is documented in the acceptance report).
- Proceeding to the next step, that is Operational Qualification, is not possible.
- Contact service department in order to have the missing element delivered (corrective action). Approval provided by the Biobank's Manager to the deviation (lack of cable) and corrective action (contact with service department).
- When the power cable is delivered, Installation Qualification is performed once again—when the result is positive, OQ and PQ are commenced.

Frequently Asked Questions
- **Is the interval between the performance of IQ, OQ, and PQ important?**
  If the time elapsed between IQ and OQ, or OQ and PQ do not influence the equipment acceptance criteria, i.e. do not evoke changes of the equipment operational parameters, it is of no importance. E.g. Installation Qualification of a centrifuge was performed with a positive result in January, and Operational

Qualification was commenced in June (due to the lack of biological material) and it is certain that no one had used the centrifuge contrary to its intended use (it is still in the same place where the service had installed it, i.e. is leveled, no elements are missing), one may proceed with Operational Qualification. If time elapsed between IQ and OQ, or OQ and PQ influenced the equipment acceptance criteria, i.e. the equipment was used with operational parameters not recommended by the manufacturer, it may influence the change of the equipment operational parameters; then, prior to proceeding with OQ and PQ, IQ should be performed once again.

7.1.2.1 Installation Qualification (IQ)

- It is required to document (Table 7.1) the performed Installation Qualification and its verification according to the indications provided by the equipment manufacturer.

Table 7.1 The minimum scope for installation qualification

Controlled elements	Acceptance limits	Result	Positive/ negative verification result
Equipment data on the equipment rating plate	Data on a rating plate should be consistent with data provided in documentation delivered with the equipment	Data are unified	P □/N □
Surrounding environmental conditions (temperature and humidity)	Surrounding environmental conditions for a given equipment should be consistent with the requirements set forth in the equipment instruction	Temperature and humidity in a room, in which the equipment is installed, are within the limits provided by the equipment manufacturer	P □/N □
Completeness of the delivery; the equipment should be unpacked from a cardboard box, all components of the equipment should be checked thoroughly	Completeness of the delivery according to the order and instruction for use	Elements present	P □/N □
Damages (checking whether the equipment was subject to mechanical damage during transport)	All equipment components are intact	No damages	P □/N □
Connection to the electric grid	Equipment is switched on when the ON/OFF switch is placed in 1 (ON) position	Works correctly	P □/N □

- In the case of transferring the equipment that has failed to the service, it is necessary to conduct the installation qualification on-site in the Biobank before using the equipment.

6.5.6 ISO 20387:2018

The Most Common Practices
- Qualified Biobank personnel with documented qualifications can carry out the installation qualification of the equipment at the target site of use at the Biobank. The qualifications of the Biobank personnel should be in accordance with the requirements of the equipment manufacturer for specific service activities. They should be documented and compatible.

Frequently Asked Questions
- **What to do if the Biobank has not received documented IQ or there was an external service delivery of the device without IQ provided?**
  Without documented IQ, we cannot accept IQ with a positive result. It is permitted for the Biobank to perform IQ. This document should be confirmed with the date and signature of the Biobank Manager or an authorized person.
- **What to do if the IQ is a negative result?**
  Identify the cause, e.g. the screen the device is equipped with does not display the image (dark screen). You cannot go to the next step: OQ. Contact a service center to repair the problem or to provide a new device or screen. The Biobank Manager acknowledges the identified deviation and corrective action (repair or replacement of the device). After repairing the device, we perform IQ—in the event of a positive result, we start OQ and PQ.

7.1.2.2 Operational Qualification (OQ)

- The Biobank should ensure that an OQ is carried out after a positive IQ result is obtained and validated.
- Operational Qualification should indicate that all operational parameters of the equipment are correct and within the acceptance limits assumed by the user.
- It is required to document Operational Qualification (Table 7.2).
- It is required that qualification tests provide for all critical elements and operational parameters of the equipment (the worst-case scenario), as set forth by the Biobank in the equipment specification (see this chapter MBQM) (based on risk assessment). In the case of a deviation during qualification and undertaking corrective actions by the Biobank, it is required to document them (see Chap. 11 MBQM).
- When the equipment failure is remedied, it is recommended to perform Operational Qualification again.
- OQ should be performed also in other critical cases that may cause the change in the equipment operational parameters, e.g. in the event of software change, when this change generates a significant impact on the device's operating parameters.

Table 7.2 The minimum scope of the Operational Qualification for centrifuge

Controlled elements	Acceptance limits	Result	Positive/ negative result
Minimum speed	100g	Value reached	P □/N □
Maximum speed	2700g	Value reached	P □/N □
Controlled temperature of centrifugation	+4 °C, +8 °C	Value reached	P □/N □
Controlled time of centrifugation process	7 minutes and 60 minutes	Value reached	P □/N □
Equipment load	Maximum load of a centrifuge with biological material	Value reached	P □/N □

- It is required to perform Operational Qualification in regular intervals during routine usage.
- It is required to define the frequency of verification of the adopted parameters for Operational Qualification during day-to-day operation of the equipment (see Sect. 7.3.1. MBQM).

6.5.12 ISO 23087:2018

The Most Common Practices
- Operational Qualification is performed by the manufacturer's service department. It is allowed for Operational Qualification to be performed by the user.
- The Biobank performs service of the device, e.g. of a laboratory centrifuge, once a year. Service activities are carried out in parameters consistent with the internal technological procedure. Documentation provided by the service engineer is equivalent with OQ confirmation.

Frequently Asked Questions
- **What is "the worst-case scenario?"**
  It is the testing of the equipment in the worst operational conditions, e.g. for a centrifuge it is a centrifugation cycle under the full load (maximum volume of biological material), at the highest rate and in the lowest temperature used in the process.

7.1.2.3 Performance Qualification (PQ)
- Process qualification shall verify that the device and its work are adequate for the process which takes place in the Biobank. Assuming specific process requirements, PQ should confirm whether the parameters in which the device works are sufficient for repetitive, optimal, and most efficient performance of individual stages of the process.

Table 7.3 The minimum scope of the laboratory equipment Performance Qualification on the example of a centrifuge

Controlled elements	Acceptance limits	Result	Positive/negative verification result
Temperature of centrifugation	+4 °C	Value reached	P □/N □
Time of centrifugation process	14 minutes	Value reached	P □/N □
Speed of centrifugation	2500 g	Value reached	P □/N □

- This is the final stage of qualification, which is performed following the positive confirmation of Operational Qualification.
- It is required to document Performance Qualification (Table 7.3).
- It is allowed to perform Performance Qualification together with Operational Qualification.
- It is allowed to perform Performance Qualification together with validation process (see Chap. 10 MBQM).
- In the case of a deviation during Performance Qualification and undertaking corrective actions, it is required to document them.
- It is required to perform Performance Qualification after modification of the assumed equipment operational parameters or with the use of other type of biological material.

6.5.5, 6.5.12 ISO 20387:2018

The Most Common Practices
- PQ is most often carried out on the final biological material (if possible) or reference material purchased from an external company or other material imitating the behavior (substitute) of biological material used in the technological process.
- The above-mentioned parameters included in PQ are identical to the parameters determined in the technological process.

Frequently Asked Questions
- **Is it allowed for the Biobank use in their process the equipment without Performance Qualification?**
  No; the equipment should be qualified to be used after Performance Qualification with the kind of biological material or certified material/substitute that will be used in the Biobank.

7.1.3 Final Qualification Report (IQ, OQ, PQ)

- It is required to document the qualification process (IQ, OQ, PQ), which should confirm that:
  - the documentation is complete;
  - qualification results are undisputed and fall within acceptance limits;
  - all deviations and corrective actions pertaining to them were confirmed.
- If all the above points are fulfilled, the result of a qualification performed in such a way is admission of a given equipment to operate.
- If one of the above points is not fulfilled, the result of a qualification performed in such a way is a conditional admission of a given equipment to operate or layoff operation.
- In any case, the decision should be justified.

6.5.7, 6.5.8 b), 6.5.8 c), 6.5.8 f), 6.5.11 the final Article of ISO 20387:2018

The Most Common Practices
- The final qualification report prepared by the Biobank often includes:
  - Name, unique designation of the equipment;
  - Dates of the performance of IQ, OQ, and PQ; it is important that the dates are subsequent to each other;
  - Recording of the documentation verification, confirming the positive performance of IQ, OQ, and PQ;
  - Recording of the deviation documentation verification, confirming positive acceptance;
  - Recording of "Equipment permitted for use";
  - Signature of the Biobank Manager, confirming qualification.
- The Biobank is equipped with two tanks intended to store biological material in liquid nitrogen vapors. Biological material is stored in one of the tanks only; the other one remains empty (backup device). The Biobank prepared equipment specification for both tanks. Both tanks are introduced to a list/register of equipment in the Biobank's equipment inventory. Both tanks were subject to successful IQ, OQ, and PQ. Owing to the performance of IQ, OQ, and PQ the additional tank is constantly ready to use (e.g., in the event of failure of the first tank).

Frequently Asked Questions
- **May the Final qualification report be an internal document of the Biobank?** *Yes; the Final qualification report may constitute an internal document of the Biobank. It is allowed to accept external qualification report, summarizing all its stages; such a document must be approved and signed by both parties, i.e. an external company and the Biobank Manager and/or a person authorized by him.*

7.2   Rules for Supervision Over Equipment

- It is required that the Biobank defines supervision for each critical equipment.
- It is allowed to adopt a type of supervision according to recommendations of the manufacturer of the equipment.
- The supervision may be divided into calibration, qualification, verification, inspection, legalization, scaling, and monitoring (see Glossary).
- It is required that the Biobank prepares a schedule of supervision over its equipment.
- It is required that the schedule is approved by the Biobank Manager or a person indicated by the Biobank Manager.
- It is required to update the schedule at specified intervals.
- The Biobank shall define the frequency of supervision performance for each critical equipment.
- It is allowed to increase the frequency of equipment supervision performance.
- The spare devices shall have the same supervision as basic (critical) Biobank equipment.

6.5.7, 6.5.10 ISO 20387:2018
7.1.5.2 a) ISO 9001:2015

The Most Common Practices
- The Biobank prepares common standard operational procedures (SOP) defining the rules for acceptance, verification, calibration, qualification, commissioning for use, operation, maintenance, supervision, and decommissioning of laboratory equipment.
- The Biobank defines the minimum scope of supervision based on the requirements defined by the device manufacturer (Table 7.4).
- The Biobank uses the following status of equipment:
  - Positive status—means that the equipment has undergone positive validation, adjustment, calibration;
  - Commence status—means the technical service should be called in;
  - Negative status—means that the equipment has no current review, calibration report, validation, or calibration;

Table 7.4 Minimum recommendations for the supervision schedule

Equipment—name, unique designation	Types of supervision	Date of supervision action performance	Frequency of supervision performance	Planned deadline of a subsequent supervision performance
E.g. laboratory centrifuge, designation 123/2017 PIK	Qualification	01 September 2017	Once a year	September 2018

- Stand by status—means that the equipment has an outdated calibration, validation, or calibration report, but is working properly (the equipment is not supervised).

- It should always be required from the manufacturer or a supplier of the equipment to define the frequency of supervision.
- If a frequency of supervision is not defined, the minimum supervision performance should take place once a year, unless the qualification process (PQ) indicates otherwise.
- In internal procedures, the Biobank determines the frequency of supervision performed both by external companies and by employees of the Biobank.

Frequently Asked Questions
- **Should the equipment without a defined supervision be used in the Biobank?**
  No, because there is no certainty that the equipment operates correctly.
- **Should a thermometer be used for the monitoring of temperature during transportation of biological material require calibration or only scaling?**
  The thermometer requires calibration in certain conditions, i.e. defining relation between the values measured and indicated with a measuring instrument and counterpart values performed based on the measurement unit templates.
- **Who defines the type of supervision?**
  The manufacturer of the equipment defines the type of supervision. The Biobank can internally determine additional supervision activities over the device that result directly from the specificity of the technological process.
- **Is it possible to extend intervals between, e.g., calibration of a thermometer used for the monitoring of temperature during transportation of biological material?**
  No; if a manufacturer clearly defined that the minimum frequency of thermometer calibration is once a year, then it means the manufacturer is able to secure its correct operation within such a period. It is always possible to address the thermometer's manufacturer, describing the way the thermometer is to be used in the biobanking process, with a request to provide any new recommendations for the operation of the thermometer that will influence the extension of the equipment calibration interval.

7.3   Recordings of Equipment Operation

- The Biobank shall provide for periodic verification of critical equipment operational parameters, e.g. prior to the commencement and end of the equipment operational usage.
- It is required to perform the recordings of verification of the equipment operational parameters and the recordings of deviations from the said parameters.

- It is required to justify the recorded deviations and document such justification (see Chap. 11 MBQM). If this is not possible, it is recommended to stop working and call the service. Until then, the device should be out of service.
- It is required for the recordings of the equipment operation (see Sect. 7.3.1, 7.3.2 and 7.3.4 MBQM) to be verified by Biobank Manager or a person indicated by the Biobank Manager at all times.

6.5.8 c) ISO 20387:2018
7.1.5.1, 7.1.5.2 a) ISO 9001:2015

The Most Common Practices

- The Biobank established in internal procedure that the verification of operation parameters of an incubator for cell cultures is performed at the beginning of a workday and prior to its end.
- The verified parameters are: temperature, humidity, and level of carbon dioxide.
- Each verification covers the recording of values of the verified parameters and the confirmation of the verification through signing the "Equipment operation chart."
- "Equipment operation chart" contains acceptance criteria for the verified parameters, e.g. reference range of the temperature for a low-temperature freezer is from $-75\,°C$ to $-85\,°C$.
- "Equipment operation chart" is located in an immediate vicinity of a given equipment, e.g. in a cabinet under the device.
- "Equipment operation chart" is presented to the Biobank Manager or a person indicated by the Biobank Manager for verification, e.g. once a month.
- The Biobank archived recordings of the equipment operation (documentation) in one place, in a way enabling easy identification, e.g. a folder with a unique documentation name.

Frequently Asked Questions

- **What is an "Equipment operation chart?"**
  Equipment operation chart is a document in which an employee records the results of the verification of the equipment in accordance with operational parameters. The frequency of the equipment operational parameters verification is defined internally by the Biobank. For example, every day, prior to the commencement of work, an employee verifies operational parameters of an incubator, i.e. reads the values referring to the temperature, humidity, and carbon dioxide level at the incubator's screen. The values read are recorded in the "Equipment operation chart"; after recording, the employee's signature is required. The signature confirms that parameters were consistent with the adopted acceptance criteria. If irregularities are identified in comparison with acceptance criteria, the Biobank Manager and the operator of the equipment should be immediately informed.
- **What to do with the equipment that is out of use, e.g. for a week?**
  When the equipment is excluded from use for a short period of time and if it is not separated from the rooms where the process is carried out, it shall be marked as "equipment out of use" on the working card and confirmed with a signature.

7.3.1   Key Parameters Defined Internally

- The Biobank should define key parameters for equipment operation during the preparation of internal equipment instruction (internal user manual).
- It is allowed to modify or extend key parameters of the equipment operation by the Biobank during Installation Qualification and Operational Qualification, which is connected with an update and re-confirmation of internal equipment instruction.
- It is required for the key parameters of the equipment operation to be approved by the Biobank Manager or a person indicated by the Biobank Manager and the approval shall be documented.

6.5.7 ISO 20387:2018
7.1.5.1 a) ISO 9001:2015

The Most Common Practices
- Except for parameters defined in point 7.3. of MBQM, it is also required to define the frequency of cleaning and disinfection of an incubator, e.g. once a week.

Frequently Asked Questions
- **May the Biobank independently change key parameters referring to the operation of the equipment?**
  Yes; the Biobank defines key parameters of the equipment operation if it confirms that the initially defined values are not sufficient and they may be extended. For instance, OQ of a tank storing the biological material in nitrogen vapors included temperature calibration inside the tank in one point, right at the tank's cover. When OQ is finished, i.e. approved, the Biobank confirms that one point of measurement is not sufficient, as biological material will be stored in the entire tank. Thus, it is necessary to perform temperature calibration at the bottom of the tank.

7.3.2   Parameters Verified/Actions Performed Directly by the User (User Maintenance)

- The Biobank should perform internal preparation of the equipment for operation according to the manufacturer's recommendations (it applies to the frequency, as well as the parameters checked and activities performed by the user of the device).
- It is required for the verification of parameters and/or actions performed directly by the users to be documented. If the manufacturer of the equipment did not define the rules for internal preparation of the equipment for operation performed by the user, they should be defined.
- The Biobank should monitor the equipment's operating parameters.

6.5.6, 6.5.7, 6.5.8 f), 8.4.1, 8.4.2 ISO 20387:2018
7.1.5.1 b), 7.1.5.2 a) ISO 9001:2015

The Most Common Practices

- The Biobank establishes the internal definition of the rules for sanitation and disinfection schedules for devices depending on their frequency of use and susceptibility to contamination. For example, in the case of a laboratory cooler where the reagents are stored, the external surface of the equipment should be disinfected once a week.
- It is a common practice to develop and implement sanitation and disinfection schedules for devices depending on their frequency of use and susceptibility to contamination. For example, in the case of a cell culture incubator, the outer surface of the device should be disinfected once a week.
- Monitoring of the equipment operation parameters can be performed manually or automatically. The Biobank defined its monitoring frequency for individual equipment operation parameters.

Frequently Asked Questions

- **May the Biobank increase the frequency of, e.g., cleaning the equipment?**
  Yes; if the Biobank employees working with a given equipment and unique biological material observe that the equipment requires higher frequency of cleaning, e.g., the manufacturer recommends cleaning the sequencer once a month, then the Biobank sets the frequency of cleaning once every two weeks.

7.3.3 Parameters Verified/Actions Performed by an External Company

- If an external verification of the equipment is required, it shall be carried out in accordance with the type of supervision adopted, at cyclic intervals (see this chapter MBQM).
- It is required that the above actions are defined in the Biobank internal procedures.
- It is required to document external verification of the equipment and archive such a documentation in a paper and/or electronic form, e.g. a scan of a document (see Chap. 3 MBQM).

  6.4.1.1 d), 6.4.1.5, 6.5.8 f), 6.5.10 ISO 20387:2018
  7.1.5.2 a) ISO 9001:2015

The Most Common Practices

- The internal instruction define parameters of the equipment. External supervision should always cover to the assumed operation parameters. If a given equipment is used to perform more than one process with different parameters, then the external supervision should consider all parameters.
- The Biobank prepares a schedule of external verification for the Biobank equipment.

- It is permissible to carry out internal supervision of equipment provided that the Biobank has the necessary competence and infrastructure.
- Internal verification of temperature sensors with externally calibrated sensors is acceptable. This means that a Biobank employee places both temperature sensors in one refrigeration unit, takes a measurement, and compares the results to the calibrated sensor. The activity should be documented.

Frequently Asked Questions
- **May the verification performed by an external supervision service include more parameters of the equipment operation than required by the Biobank?**
  The external supervision service must, first and foremost, verify the parameters of the equipment operation that are crucial for the biobanking process. It is permitted to extend those parameters during actions performed by the external supervision service.

7.3.4   Recordings From Maintenance, Cleaning of the Equipment

- The Biobank should define cleaning and disinfection methods for critical equipment.
- It is required to define frequency of performing cleaning and disinfection of the equipment.
- It is required to document the performance of cleaning and disinfection of the equipment.
- It is required for the Biobank Manager or a person indicated by the Biobank Manager to verify the documented cleaning and disinfection of the equipment.

7.1.5.1 last paragraph ISO 9001:2015

The Most Common Practices
- Most often, the Biobanks use one sheet of recordings of the equipment operation which include calibration, cleaning, and disinfection.
- Most often, the Biobanks in internal procedures determine the frequency of maintenance, cleaning, and disinfection based on the manufacturer's recommendations.
- Most frequently, the Biobanks archive the recordings of maintenance, cleaning, disinfection of the equipment in one place, in a way enabling an easy identification, e.g. a folder with a unique equipment name.

Frequently Asked Questions
- **May the Biobank internally define the frequency of the verification of the effectiveness of cleaning and disinfection of the equipment?**
  Yes; the Biobank should define the frequency, as well as the method of performing the verification of the effectiveness of cleaning and disinfection of the equipment. Frequency should be adequate to the sterility required in a given process. For

instance, the laminar hood should be checked for the presence of mycoplasma, this requirement is in line with technological process, then the Biobank establishes the criteria for cleaning and disinfecting the laminar chamber itself.

7.4 Work with Equipment Outside the Supervision

- It is required to define in the Biobank's internal procedures of work with the equipment which operation parameters are not met. The following should be defined:
  - the way of labeling the equipment which should be unique and unmistakable;
  - the manner in which all employees of the Biobank are notified about the decommissioning of the equipment;
  - if possible, the equipment which is not subject to supervisions should be separated from the equipment operated during a day-to-day work.

6.5.11, 6.5.12 ISO 20387:2018
7.1.5.2 c) and last paragraph ISO 9001:2015

The Most Common Practices
- A device that has been identified as not meeting the operating criteria is marked with a label: "CAUTION, device out of service."
- Separate the equipment and secure it against unintentional usage, e.g. seal the incubator's doors with lead if the microbiological inoculation result is positive. Record the failure information in the device documentation, e.g., in the incident sheet.

Frequently Asked Questions
- **What does it mean that the device does not meet reference operating parameters?**
  It is a device which after the verification performed by the employee (e.g., before starting work) does not meet the internal criteria set by the Biobank. e.g. the fridge in which the reagents are stored should work in the following temperature range: from +2 °C to +8 °C. When the Biobank employee after coming to work verifies the temperature and identifies that it is higher than the accepted reference range, i.e. +24 °C, then the employee transfers reagents to another refrigerator and the device is labeled "CAUTION, device out of service." By email, he sends information to other Biobank employees that fridge No. 00123 is being put out of service until the failure is removed.
- **What to do if the equipment that is not subject to supervision was used in the process?**
  If the equipment outside specification was used in the process, neither the obtained result nor the safety of biological material may be considered certain. The results should be repeated using another controlled device. E.g.: the Biobank employee places test tubes with biological material in a centrifuge, starts the

program, checks whether the program works properly. The spin process is running correctly. The Biobank employee goes to another room to enter data into the computer. The spin process is complete, the device signals the end of work. The employee opens the centrifuge and takes out samples. During this operation, he observed that the samples did not look right, so the centrifugation process did not proceed according to accepted internal rules. In such a case, follow the guidelines of the procedure describing the procedure after the detection of noncompliance (see Sect. 11.2 MBQM).

7.5   Requalification of Critical Equipment

- Requalification is a periodic process that is a continuation of OQ and/or PQ in everyday usage of the equipment (see Sect. 7.2 MBQM).
- The Biobank should provide for the requalification of critical equipment.
- The performed requalification process should be documented.
- The frequency of requalification should be defined in the supervision schedule (see Sect. 7.2 MBQM).
- The requalification process should secure the following:
  - traceability of the objective of the requalification process;
  - description of the expected results;
  - acceptance limits for the process;
  - approval of supervisors pertaining to the above three points;
  - overview of the obtained results with assumptions of the three initial points and conclusions drawn;
  - approval of supervisors for the obtained results and conclusions preparation;
  - it is required to document the conduct if process acceptance criteria are not met;
  - it is indispensable to obtain a final approval for the requalification from THE Biobank Manager or a designated person;
  - the Biobank employees training—if required.

6.5.8 ISO 20387:2018
8.5.1 f) ISO 9001:2015

The Most Common Practices
- Requalification of an incubator for cell cultures is performed once a year by an external company. Requalification is to confirm the correct maintenance of the following incubator's operation parameters: temperature, humidity, and carbon dioxide level. The Biobank Manager approves of the acceptance limits. An external service company performs requalification within the adopted acceptance limits, and the action is confirmed through the delivery of requalification documentation. The Biobank Manager verifies the delivered documentation and approves of the positive result of the requalification process.

Table 7.5 Equipment requirements

Requirement	Critical equipment	Other equipment
Procedure for acceptance of the equipment	REQ	REQ
Procedure for qualification of the equipment	REQ	REQ
Supervision over the equipment	REQ	REC
Equipment specification	REQ	REC

REQ required, *REC* recommended

- In the case where the Biobank has a qualified equipment to perform the requalification of such an equipment, it can carry out this process itself. The Biobank Manager verifies the requalification documentation and accepts the positive result of the requalification process (Table 7.3).
- The Biobank requirements for critical and other equipment (Table 7.5).

Frequently Asked Questions
- **Should the Biobank perform the requalification of the equipment?**
  Yes; requalification is an action that aims at confirmation and documentation (proof) that a process performed within the set range of parameters is performed effectively and in a repeatable way, and fulfills the set requirements concerning a specification and qualitative criteria. Requalification confirms the correct operation of the equipment, thus guarantees the quality of biological material and provides for its safety.

References

ISO 20387:2018 Biotechnology – Biobanking – General Requirements for Biobanking.
ISO 9001:2015 Quality management systems - Requirements.

Traceability

8

Abstract

Traceability in biobanking guarantees the quality of biological material and related data. The traceability system implemented in the Biobank must ensure the bi-directional (two-way) traceability of both material and data, as well as each sample generated from such material and data. The lack of the bi-directional traceability system significantly reduces the value of biological material and the reliability of the results of future analyses. One way to ensure traceability is to introduce a unique identification of biological material and related data. This chapter describes the system that enables the full identification of biological material and related data received in and issued from the Biobank. It also applies to the unique labeling of biological material and data, which allows them to be identified at every stage of their life cycles.

8.1 Traceability System

- The Biobank shall establish, implement, and use a system ensuring bi-directional traceability of biological material and associated (related) data accepted to (received in) and released (issued) from the Biobank.
- The Biobank shall ensure the ability to locate biological material and associated data by creating and storing documented information (records, biobanking protocols) documenting the course of individual stages of critical processes of handling biological material.
- The traceability system shall ensure that the documented information created in the Biobank (records, protocols from the biobanking process):
  - arises in real time, i.e. in parallel with the activities performed or as soon as possible, e.g. immediately after the completion of the activity;
  - is unambiguously and permanently associated with biological material;
  - allows the identification of:

A. Matera-Witkiewicz et al., *Manual of Biobank Quality Management*,
https://doi.org/10.1007/978-3-031-12559-1_8

(a) the person who is directly responsible for the activity;

(b) materials and reagents used;

(c) devices used;

(d) places (e.g., rooms) where the activity was carried out;

(e) the date and time of the beginning and end of the activity (in the standard format specified in ISO 8601: 2004);

- confirms that the activities were carried out in accordance with the relevant (i.e., taking into account the given type of biological material and its possible purpose) versions of the applicable procedures for a given technological process or they record deviations, discrepancies, and deviations;

- is accessible, readable, durable, and protected against unauthorized changes.

• The traceability system shall ensure that all biological material and related data are associated with the records of details of permits or restrictions associated with their use.

7.5.1 ISO 20387:2018
8.5.2, 8.6 b) ISO 9001:2015

The Most Common Practices

• Ensuring a bi-directional traceability of biological material is possible by simultaneously making the following possible:

- (1) the ability to identify and locate biological material and associated data at any stage of the technological process, in particular at the stage of:

(a) obtaining (e.g., collection/procurement) material and data by the Biobank for the biobanking;

(b) transport of material and data to the Biobank;

(c) receiving material and data in the Biobank;

(d) qualification of received material and data in the Biobank;

(e) acceptance of qualified material and data to the Biobank;

(f) processing of accepted material and data in the Biobank, e.g. during division into samples (if applicable);

(g) storage of material and data in the Biobank;

(h) distribution (release, issue) of material and data from the Biobank to the recipient/user or its utilization, including the ability to identify and locate:

i. current versions of relevant procedures related to critical processes at the Biobank, including procedures for obtaining, transport, receiving, qualification, acceptance, processing, storage, and distribution (release, issue) of biological material and associated data;

ii. documented information (records, protocols from the biobanking process) regarding the course of the above-mentioned critical processes;

- (2) the ability to identify:

(a) the donor of biological material/source of the data;

(b) the Biobank personnel, in particular persons performing the activities during critical processes and individuals supervising the course of these processes and verifying compliance with the acceptance criteria;

(c) materials, in particular critical materials used for the activities performed during critical processes;

(d) equipment, in particular critical equipment used for the activities performed during critical processes;

(e) rooms in which the activities are performed during critical processes;

(f) entities involved in the handling of biological material, including external entities cooperating with the Biobank, e.g., as part of outsourcing contracts or recipients and persons in the above-mentioned entities involved in critical processes;

– (3) the ability to locate all relevant information/data pertaining to persons/ entities, items/objects referred to (2), i.e., the ability to locate documented information regarding:

(a) the sources of biological material/data, in particular documented information about the donor's will regarding a specific donation of biological material (ability to locate the donor's consent form, if applicable) and procurement of biological material form the donor (ability to locate biological material procurement report);

(b) the Biobank personnel (ability to locate, i.a., the documents confirming their qualifications and professional experience as well as the documents confirming completion of required training);

(c) materials (ability to locate, i.a., quality certificates for critical materials);

(d) equipment (ability to locate, i.a., equipment cards and documentation regarding the qualifications of critical equipment);

(e) premises (ability to locate, i.a., documentation for room cleaning and disinfection);

(f) external entities cooperating with the Biobank (ability to locate, i.a., the outsourcing contracts).

• When entering data into a computer database, it is desirable to check the data for compliance with the requirements (e.g., database requirements or user requirements) in real time. For example, if a person enters a date format that does not comply with the requirements, the person is notified and must correct the error.

Frequently Asked Questions

• **What does the bi-directional traceability system mean?**

It is a system that makes it possible to trace the history (fate) of biological material and associated data from their donor to their recipient and from their recipient to their donor.

• **Does the need to ensure the bi-directional traceability of biological material and associated data also mean the need to ensure bi-directional traceability of each sample prepared from biological material and associated data?**

Yes, the Biobank traceability system implemented in the Biobank must ensure the bi-directional traceability of both biological material and related data as well as each sample generated from this material and related data.

- **Why is it important that records (protocols from the biobanking process) created in the Biobank be clearly and permanently linked to biological material?**
  This is important because the lack of detailed information linked to biological material significantly limits its value and the reliability of the results of future analyses.

8.2   Unique Identification of Biological Material

- The Biobank shall establish, implement, and use a procedure for assigning biological material and associated data unique identification, which shall identify material and data at every stage of their life cycle.
- The content of the above-mentioned procedure shall take into account applicable legal and environmental requirements regarding the labeling of biological material and associated data, as well as consider all storage conditions of individual types of biological material implemented in a given Biobank, e.g. storage temperature and type of containers used.
- The Biobank shall ensure the ability to identify biological material and associated data by assigning them a unique identification.
- The uniqueness of identification of biological material and associated data must be ensured at least at a given Biobank.
- The unique identification is assigned by the Biobank upon acceptance of the qualified biological material and associated data.
- The unique identification assigned to biological material must be permanent regardless of the time and conditions of storage of biological material and associated data in the Biobank.

7.5.1 a) ISO 20387:2018

The Most Common Practices
- Unique identification of biological material is usually created by using the combination of several identifiers, i.e. several alphanumeric sequences.
- Determining the rules for creating identifiers enables the introduction of a coding system in the Biobank. The introduction of the coding system enables the introduction of graphical representations of information, e.g., in the form of bar codes or 2D codes.
- The rules for creating identifiers and the order in which they should be located in the unique identification should be established and described in detail in the relevant procedure.
- For the unique identification of biological material and associated data, a combination of the following identifiers is most often used:
  - an identifier identifying a specific Biobank;
  - an identifier identifying a specific project;

- an identifier identifying the date (year, month, and day) of the specific donation of material and data;
- an identifier identifying a specific donation of material and data in a certain
  year, month, and on a certain day;
- an identifier identifying a specific sample of biological material resulting from
  the processing of specific material.
- An example of unique identification: *BWP10120033113589*, where:
  - *BW* is an identifier identifying Wroclaw Biobank;
  - *P101* is the identifier identifying a specific project in Wroclaw Biobank;
  - *20* is the identifier of the year of the specific donation of material and data, *03*
    is the identifier of the month of the specific donation, and *31* is the identifier of
    the day of the specific donation;
  - *13* is the identifier of a specific donation in a specific year, month, and day
    (31.30.2020) in a specific project (P101), in a specific Biobank (*Wroclaw
    Biobank*);
  - *58* is the identifier of a specific sample of specific biological material, e.g.
    a pipetted cryotube.
- The unique identification is placed on containers with biological material (e.g., on
  the container in which biological material was transported to the Biobank, on
  containers such as cryotubes with material samples) and in the documentation
  available at the Biobank, including in particular:
  - the material and data donor consent form;
  - the collection (procurement) and transport protocol/report;
  - all documents regarding material and data handling at the Biobank, including
    in the following documents:
    (a) the reception, qualification, and acceptance report;
    (b) the processing report;
    (c) the storage report;
    (d) the distribution (release) and transport report;
    (e) the utilization report.
- As carriers of unique identification, usually printed labels permanently glued to
  documents and containers with biological material (e.g., tubes, slides, boxes) are
  used most often.
- Verification of data entered into the computer database in terms of compliance
  with the requirements (database, user, etc.) takes place in real time. For example,
  the computer database checks whether the format of the date entered is correct. If
  not, the user is informed and must correct the error.
- For example, protocols/reports are created in the case of an intentional (carried
  out directly to acquire biological material for the Biobank), instrumental
  (performed with the use of the equipment/tools such as a venous blood collection
  kit, biopsy needle/trepane, endoscope), invasive (e.g., associated with the disrup-
  tion of integument, use an endoscope) collection of biological material performed
  on behalf of the Biobank from an adult donor and subsequent securement of the
  collected material.

- Forms/prints of such protocols/reports usually contain the following fields for providing information on (see also Most common practices in subchapter 10.1.1 MBQM):
  - collection protocol number (submitted by the Biobank);
  - data regarding the research project:
    - (a) designation and name/title of the research project;
    - (b) designation and date of the document issued by the Bioethics Committee and approving the research project protocol;
    - (c) identification of the main researcher: name and surname, professional title, contact details, including current telephone number;
    - (d) designation and version number/edition/issue of document containing information for the donor about the research project, including information on:
      - i. the method of biological material collection and related possible side effects / adverse events or reactions;
      - ii. the provision of care to the donor after collection;
      - iii. further handling of collected biological material and related data;
      - iv. fesignation and version/edition/issue number of the donor informed consent form/print for the collection of biological material for a specific purpose, for further specific handling of this material in the Biobank (i.a. for its processing, storage and testing) as well as for the use of related data by the Biobank;
  - data regarding QMS:
    - (a) designation and version/edition/issue number of the procedure/instruction according to which biological material has been collected;
    - (b) designation and date of concluding a contract between the Biobank and the medical entity for the collection of biological material and the provision of care to the donor after the collection;
  - data regarding the donor:
    - (a) donor identification information: name and surname, date of birth/ personal identification number; contact details (necessary in the event of subsequent complications);
    - (b) information on enclosing or not enclosing an appropriate form/print of the donor informed consent to the collection protocol/report and in the case of its enclosure—confirmation that it has been filled correctly by the donor and by the person obtaining this consent or specification of the discrepancies found;
    - (c) information enclosing or not enclosing an appropriate form/print of donor medical/health questionnaire and in the case of its enclosure—confirmation of the correctness of its completion or specification of the discrepancies found;
    - (d) identification data of a person performing a check of the donor's identification: name and surname, position as well as the signature of this person;

(e) unique designation of the donor's documentation in the medical entity,[1] e.g. hospital main register number, medical record number, number in blood donation register, number in the register of endoscopic examinations;

- data regarding the place/location of collection:
  (a) identification data of the medical entity where the collection took place: name and contact details;
  (b) name of the department / unit/ ward of medical entity (e.g., collection point/unit, surgical office, endoscopic room, operating theatre) and its contact details, if different from the data of the medical entity;
- data regarding the collection:
  (a) identification data of a person collecting biological material: name and surname, professional title, position as well as the signature of this person;
  (b) confirmation by the person collecting biological material of his/her knowledge and compliance with the applicable version/edition/issue of procedure/instruction for collecting and securing of the collected biological material: signature of this person;
  (c) date and time of collection of biological material (in a format compliant with the ISO 8601 standard) or, in the case of longer procedures, date and time of the beginning and the end of the collection of biological material (in a format compliant with the ISO 8601 standard);
  (d) information on the course of the collection and the occurrence/absence of complications/adverse reactions during the collection;
- data regarding biological material collected and related data:
  (a) identification data of biological material:
    i. type of material collected (e.g., whole venous blood, section of skin, gastric mucosa, bronchoalveolar lavage fluid);
    ii. size of material collected: mass [g] (if applicable), volume [ml] (if applicable);
    iii. number of containers with material collected [items];
    iv. method/methods of securing the material collected (e.g., making a smear on a slide with subsequent fixation, placing material in a transport medium, placing in a fixative (e.g., formaldehyde, glutaraldehyde), cooling from +2 °C to +8 °C, freezing, providing protection against access to light);
  (b) identification data pertaining to products and materials in direct contact with the biological material collected:
    i. type, lot number, expiration date, aseptic/sterile status (aseptic/sterile, non-aseptic/non-sterile), manufacturer: e.g., container details, fixative medium details, transport medium details;

---

[1] Procurement (collection) of biological material, which is an intentional and/or instrumental and/or invasive procurement (collection), can only take place in a medica entity.

(c) identification of a person responsible for securing the material collected along with related data until the time it is transferred to the Biobank: name, position, and signature of the person.
- The form/print of the collection protocol/report is often combined into one whole with the form/print of the protocol/report of:
  - the delivery (transport) of collected, secured biological material, and associated data to the Biobank;
  - the reception of delivered biological material in the Biobank;
  - the qualification of receipted biological material, ending with either acceptance, labeling (giving a unique identifier) and transferring of qualified material for processing and/or storage in the Biobank or utilization of unqualified material (in the case of its failure to meet the set requirements/acceptance criteria).

Frequently Asked Questions
- **Does the Biobank have to create unique identification each time?**
  The Biobank must use unique identification, however, it can use, for example, ready-made cryotube manufacturer codes or use the codes that have been proposed in a specific project, provided they meet the condition of uniqueness. The adoption of the above-mentioned solutions must be reflected in the relevant procedure.
- **What are the advantages of introducing graphical representations of information, e.g. in the form of bar codes or 2D codes?**
  The introduction of the above code allows the automation of information transmission through the use of code scanners. The use of scanners, in turn, speeds up the flow of information and reduces the risk of error.

References

ISO 20387:2018 Biotechnology—Biobanking—General Requirements for Biobanking.
ISO 9001:2015 Quality management systems—Requirements.

Environmental and Staff Hygiene

9

Abstract

The rules related to ensuring a proper work environment for employees and the relevant rules regarding the environment of the processes taking place in the Biobank directly affect the quality of biological material and/or data. In this chapter, attention is drawn to the requirements concerning occupational health and safety, principles of handling chemical and hazardous substances, potentially infectious material, and personal protective equipment. We also outlined the rules for work involving biological material (current medical certificate confirming the ability to work as part of preventive examinations and vaccinations) and the procedures to be followed in the event of an employee's exposure to potentially infectious material.

Attention was drawn to the fact that the Biobank should enable the processes to be carried out in an environment that does not have a negative impact on their course. The chapter describes the principles of keeping the facilities clean and the issues related to the management of waste, including hazardous waste. Moreover, sanitary plans in force in laboratories, which usually contain information on surface and equipment disinfection and sanitisation methods and guidelines for the preparation of cleaning and disinfecting agents, were discussed.

9.1 Personnel Hygiene and Safety

- The Biobank shall develop, implement, and apply procedures describing the requirements of occupational health and safety, rules of dealing with chemical and hazardous substances, potentially infectious material and personal protection.

4.1.1, 4.1.7, 6.2.1.5, 6.2.3.1, 6.3.1, 6.3.5 ISO 20387:2018
7.1.4 ISO 9001:2015

© Wroclaw Medical University 2023
A. Matera-Witkiewicz et al., *Manual of Biobank Quality Management*,
https://doi.org/10.1007/978-3-031-12559-1_9

9.1.1   Occupational Health and Safety

- The Biobank shall develop the written guidelines for safe work at the Biobank, taking into account chemical agents and reagents as well as biological material.
- These guidelines shall be regularly reviewed, updated, and modified (in response to problems or if the actions adopted so far prove to be ineffective).

4.1.1, 6.2.1.5, 6.3.1., 6.3.2, 6.3.5 ISO 20387:2018

9.1.1.1 Occupational Risk
- In the case of activities that may cause a risk of exposure to biological agents, the Biobank shall implement procedures to determine the type, degree, and duration of the exposure of employees to identify the necessary precautions.

6.3.5 ISO 20387:2018

The Most Common Practices
- The risk assessment of personnel at the Biobank should be carried out at regular intervals and whenever changes in working conditions that may affect the exposure of workers to biological agents are made. Occupational risk assessment is handled by health and safety inspectorates or relevant organizational sections of the unit.
- Among the factors affecting the assessment of occupational risk, the following can be enumerated: biological, chemical, physical, and psychophysical factors.

9.1.1.2 Periodic Examinations
- The Biobank or the unit of which it is a part should develop, implement, and apply a health and safety training program.
- Each employee shall undergo training in security areas appropriate for his workplace.
- The training shall be updated at regular intervals or repeated if clearly needed.
- Training for the Biobank staff shall also include emergency procedures for the workplace.

6.2.1.3, 6.2.1.5, 6.2.2, 6.2.3, 6.3.7, 7.7.1 ISO 20387:2018

The Most Common Practices
- Periodic examinations should include information on: potential health risks; precautions to be taken to minimize exposure; hygiene requirements, clothing, and using personal protective clothes and other protective equipment; steps employees should take in the event of accidents at work or in order to prevent accidents.

9.1.1.3 Personal Protective Equipment
- The Biobank should provide employees with:
  - protective or work clothing and ensure its proper use and maintenance;
  - personal protective equipment;
  - gloves, masks, caps.

5.4, 5.6, 6.1.1, 6.3.1, 6.3.4 ISO 20387:2018
8.1 b1) ISO 9001:2015

The Most Common Practices
- Protective equipment (aprons, footwear, face shield, gloves) is stored in a clearly marked place (e.g., described container, designated place on the shelf, hanger), checked, cleaned, regularly renewed or replaced, in accordance with the procedures adopted at the Biobank.
- Eyes and mucous membranes must be protected from contact with biological substances and chemicals. Depending on the probability of exposure, goggles, safety goggles, or face shields may be used for protection.

9.1.1.4 Handling of Clothing and Protective Equipment
- The Biobank shall develop the written rules for the handling of clothing and personal protective equipment.
- The Biobank shall specify the rules for the use of protective and work clothing by employees and visitors.
- The Biobank shall provide employees with the opportunity to store private and work clothing separately.
- Protective clothing shall be removed after finishing work and stored in a designated place.
- Clothing used to work with biological material shall not leave the designated work area and it is also advisable to store it without contact with private clothing and clean clothing.
- The Biobank shall ensure that clothing and protective equipment are regularly sanitized and cleaned or disposed of as necessary.

6.2.1.5 ISO 20387:2018
8.1 b1) ISO 9001:2015

The Most Common Practices
- All persons, including visitors to the Biobank, should wear appropriate protective clothing (i.e., lab coats, long and covering trousers).
- Using clothing in different colors to different laboratory zones makes it easier to control compliance with the rules of working clothes.
- The Biobank signs a contract with the laundry for cleaning and disinfection of protective clothing for workers which have a contact with biological material.

Frequently Asked Questions
- **Is it permitted for employees to wash protective clothing at home?**
  It is not allowed to wash protective clothing, which has had potential contact with biological material, at home. Such activities may lead to the infection being transferred from the Biobank to the outside.

9.1.2   Chemical and Hazardous Substances

- All chemical and hazardous substances used in the Biobank shall have material safety data sheets.
- These data sheets shall be made available to employees, periodically verified and updated if necessary.
- The Biobank shall develop an inventory of hazardous substances used in technological processes and provide security measures, appropriate storage and disposal conditions for these substances listed in their safety data sheets.
- Work with chemicals shall be described in the appropriate procedures and operating instructions.
- Chemical substances shall be stored in original packaging, provided with original labels. It is allowed to transfer a chemical (also hazardous) from the original packaging to another packaging, provided that the following conditions are met:
  - replacement packaging should ensure a level of safety of use at least as high as the original packaging;
  - the unoriginal label should include the unique name of the chemical, its expiry date and LOT number;
  - materials and reagents should be stored in the warehouse in accordance with the manufacturer's instructions (see Sect. 6.4 MBQM).

**6.2.1.4, 6.2.1.5, 6.3.1, 6.3.2, 6.3.4, 6.3.7 ISO 20387:2018
8.1 b1) ISO 9001:2015**

The Most Common Practices
- It is recommended to implement the list of safety data sheets for substances used in the Biobank. The list shall be updated and presented to employees so that they can read it.
- Procedures describing work with chemical substances shall take into account the following:
  - the manner of protecting an employee while working with a given substance (personal protective equipment, fume hoods);
  - ways to prevent chemical leaks;
  - the manner of removing waste and other chemically contaminated materials;
  - the mode of proceeding in case of direct exposure of a worker to a chemical.
- Chemicals warehouses may be equipped with: devices providing hazard signaling (e.g., fire alarm system), appropriate equipment and extinguishing media, means for collecting the released substance (sorbents, plastic bags, closed chemical resistant containers), first aid kit, and personal protective equipment.

- If in the Biobank there is exposure to carcinogens or mutagens, the Biobank Manager is obliged to keep strict records of the following:
  - the exposure time on one working day;
  - the amount of substances used during a work shift;
  - concentrations of carcinogenic or mutagenic substances in the workplace.
- Substances/preparations are classified in terms of the following:
  - threats to human health and the environment;
  - threats resulting from physicochemical properties.
- In rooms where installations are installed that pose a risk of reducing the amount of oxygen in the atmospheric air (e.g., liquid nitrogen installations), it is recommended to install appropriate monitoring devices, e.g. oxygen and carbon dioxide (CO_2) sensors. In addition, the rooms should be equipped with effective ventilation as well as appropriate protective clothing (e.g., gloves, face and body protection).
- People working with dry ice in the laboratory should wear suitable protective clothing approved for low temperatures. Prolonged exposure to dry ice can cause severe skin damage.

9.1.3 Work with Potentially Infectious Material Derived from Humans

- Any biological material (derived from humans) shall be treated as potentially infectious.
- Biobank shall provide its employees with appropriate personal protective equipment and means for safe work with such material.
- Before starting work with biological material, each employee shall have a current medical certificate stating the ability to work as part of preventive examinations and preventive vaccinations.
- Every Biobank employee who may come into contact with blood or other potentially infectious materials shall be trained in the handling of biological material, the use of personal protective equipment and disinfection methods. The training should be documented.

6.2.1.5 ISO 20387:2018
8.1 b1) ISO 9001:2015

9.1.3.1 The List of Employees Exposed to Biological Agents
- The Biobank shall keep a register of works exposing workers to a harmful biological agent classified in hazard class 3 or 4, and a register of workers exposed to a harmful biological agent classified as hazard class 3 or 4 (in accordance with the relevant law).

6.2.1.5, 6.3.1, 6.3.4, 6.3.7 ISO 20387:2018
8.1 b1) ISO 9001:2015

The Most Common Practices
- If the Biobank works with biological, potentially infectious material, additional information may be added in the evidence list of employees exposed to infection, such as: type of work or activities performed, type of biological agent, cases of infection.

9.1.3.2 Occupational Exposure
- The Biobank or a unit of which it is a part, shall develop, implement, and apply procedures based on the risk of the employees being exposed to potentially infectious material.
- All occupational exposure events shall be recorded in an appropriate register, and their reporting should be in a documented form (in written or by electronic means).

6.2.1.5, 6.3.1, 6.3.4, 6.3.7 ISO 20387:2018
8.1 b1) ISO 9001:2015

The Most Common Practices
- Implementation of the *Occupational exposure form print,* in which the description of the situation, place and type of exposure, conditions (e.g., during intravenous injection of the right forefinger with a needle with blood through a glove) are presented, witnesses of the situation, behavior after exposure are recorded.
- Occupational exposure register includes:
  - subsequent incident number in the current year;
  - date and time of an incident;
  - name and surname of the person exposed;
  - employee's serological status (e.g., vaccination against hepatitis B in year . . .);
  - detailed description of an incident: place and type of exposure, situation in which exposure took place;
  - name and surname of a patient, his serological status (if known);
  - description of activities performed after exposure;
  - signature of exposed person and the Biobank Manager.

9.1.3.3 Disinfection
- The Biobank shall establish a procedure for decontaminating and disinfecting rooms, surfaces, equipment (see Sect. 9.2.1 MBQM) as well as skin and mucous membranes after exposure to a biological agent.

6.3.4 ISO 20387:2018
7.1.4 ISO 9001:2015

9.2 Process Environment

- The Biobank shall ensure that processes are conducted in an environment that does not adversely affect their course.

4.1.1, 6.1.1, 6.3.2, 6.3.4, 6.3.5, 6.3.6, 6.3.7 ISO 20387:2018
7.1.4 ISO 9001:2015

9.2.1 Environmental Conditions and Staff Hygiene

- The Biobank shall develop, implement, and apply a lab/room hygiene procedure.
- Sanitary operations shall be documented.
- The Biobank shall develop sanitary plans for laboratories, and it is allowed for a group of rooms/labs having similar processes to have one hygiene plan. Hygiene plans contain information on how to disinfect and sanitize surfaces and the equipment, as well as guidelines for the preparation of cleaning and disinfecting agents.
- The room hygiene procedure shall contain information about the supervision over cleaning and disinfection activities, as well as indicate the responsibilities in this regard.
- The procedure shall include information on schedules and frequencies of activities, selected sanitization and disinfection techniques, equipment as well as actions taken in the event of an emergency. When outsourcing sanitization and disinfection activities to an external company, it is recommended to create documented information on the principles of supervision over the work of persons who are not a personnel of the Biobank.
- Persons performing sanitization and disinfection of premises shall be trained in accordance with the guidelines of the Biobank hygiene procedure (e.g., through internal training).
- In a situation when the Biobank outsources sanitization activities to external companies, it shall submit for approval the documented information on the requirements and acceptance criteria and supervision of the outsourced activities.

4.1.5, 6.2.3.1, 6.3, 6.3.5, 6.4.1.1, 6.4.1.2 ISO 20387:2018
7.1.3, 7.1.4 ISO 9001:2015

The Most Common Practices
- The hygiene plan may be prepared in the form of a table specifying:
  - washing/disinfection object (e.g., small reusable equipment—e.g. pipette, small surfaces—e.g. worktops, large surfaces—e.g. floors, walls);
  - activity (e.g., disinfection by spraying or wiping, sanitation);
  - agent (providing the name and concentration of the disinfectant, washing or washing-disinfecting agent used in the Biobank);

- the frequency of performance (e.g., before and after work, after each day of work, once a week, if needed);
- the person responsible for the activity (signature of the person performing the activity).
• Hygiene plans also contain information about areas where eating, drinking, storing food and drink, smoking, chewing gum, and cosmetics are prohibited.

9.3   Waste Management

• The Biobank shall develop the rules for segregation, marking/labeling, and storage of generated waste.
• It shall provide a suitable place for storing waste until it is collected by a specialized company authorized to collect, transport, and utilize waste in accordance with applicable regulations.
• Storage and disposal of waste shall take place in a way that does not affect the proper course of the biobanking of biological material. This also applies to ensuring proper rules for the collection, storage, and disposal of waste as well as procedures for their disposal and transport.

6.3.1, 6.3.3, 6.3.5, 6.4.1.4, 6.4.1.2 ISO 20387:2018
4.1 ISO 9001:2015

The Most Common Practices
• Labeling of waste bags and hard-walled containers for waste is carried out in accordance with the requirements stipulated by the legal provisions of a country where the Biobank's premises are providing the address and name of the waste producer, waste generated code, the place of its manufacture, the date of opening the bag and the date of closing the waste bag with the signature of the closing person; opening date means the date the first waste was thrown into the bag.
• All containers used for storing waste must be hermetic and must not show any signs of damage.
• Liquid waste containers are filled up to 70–80% capacity to minimize the risk of leakage.
• Containers used should be appropriate to the type of waste stored.
• Waste is stored in a designated space, properly described and with limited access for unauthorized persons.
• It is unacceptable to open once closed waste containers or disposable bags.
• If the bag or container is damaged, it should be placed in another larger, undamaged container or bag.

Frequently Asked Questions
• **Is the "star" next to the waste code an integral part of it and should it be placed with the waste designation?**

YES, the "star" symbol is an integral part of the waste code and must not be omitted when labeling containers.

References

ISO 20387:2018 Biotechnology—Biobanking—General Requirements for Biobanking.
ISO 9001:2015 Quality management systems—Requirements.

Biobanking Processes and Quality Control 10

Abstract

The knowledge of individual stages of the life cycle of specific biological material and related data is crucial for the proper functioning of the Biobank. It helps to define and verify all important technological processes which are carried out in relation to both materials and data. The knowledge of technological processes, on the other hand, allows the Biobank to develop and implement appropriate quality control methods. Owing to the appropriate technological processes and quality control, biological material and related data are valuable and useful, and the results of the analyses performed are reliable. This chapter describes the life cycle of biological material, taking into account critical technological processes, including the collection, transport, reception, qualification, acceptance, processing, storage, and distribution of biological material and related data, as well as the quality control process.

10.1 The Life Cycle of Biological Material and Critical Processes

- The Biobank shall establish, implement, and use detailed technological procedures tailored to each biological material (both unprocessed material and processed material, i.e. samples of biological material) and related (associated) data in a specific process of a specific life cycle.
- The Biobank shall identify the stages of the life cycle of each biological material and associated data, as well as define and verify the relevant technological processes to which they are subject to. The life cycle of biological material and associated data in the Biobank depends on their type and final destination.
- The Biobank shall establish a workflow, which will describe the identified stages of the above-mentioned life cycle as well as procedures for each relevant process (see Chap. 3 MBQM).

- The Biobank shall identify and document all critical activities within each procedure of the specific life cycle process.
- The Biobank shall ensure that all procedures are up to date and easily accessible to staff.
- The Biobank shall ensure that for each biological material both the dates of the critical life cycle stages and their duration are documented using the standard format specified in ISO 8601:2004.

7.1.1, 7.1.2, 7.1.3, 7.8.2.7 ISO 20387:2018
7.5.2 c), 8.4.1, 8.5.1 c) ISO 9001:2015

The Most Common Practices
- Important technological processes at the Biobank are usually:
  - the process by which the Biobank acquires biological material and associated data, e.g. by:
    (a) the collection (procurement) of material by the Biobank or under the responsibility of the Biobank;
    (b) the transfer of material and data to the Biobank, e.g., based on a cooperation agreement. In such a case, the Biobank is not responsible for the procurement of the material;
    (c) taking responsibility for the supervision over the material and data entrusted to the Biobank as well as related data. In such a case, the Biobank is not the owner of the material and data;
  - the process of transport of biological material and associated data to the Biobank;
  - the process of reception, qualification, and either of acceptance of biological material and associated data for the inventory of the Biobank or their disposal (utilization);
  - the process of handling of biological material and associated data in the Biobank, including:
    (a) the process of processing of material and data;
    (b) the process of storage of material and data;
    (c) the process of transport of material and data within the Biobank;
  - the process of withdrawing of biological material and associated data from the Biobank by:
    (a) the process of distribution (issuing) of material and data as well as the process of their transport from the Biobank;
    (b) the process of the disposal of material and data.

The example of the general life cycle of biological material and the examples of critical technological processes are presented in Fig. 10.1.

Caution: The life cycle of biological material and related data in the Biobank depends on their type and final destination.

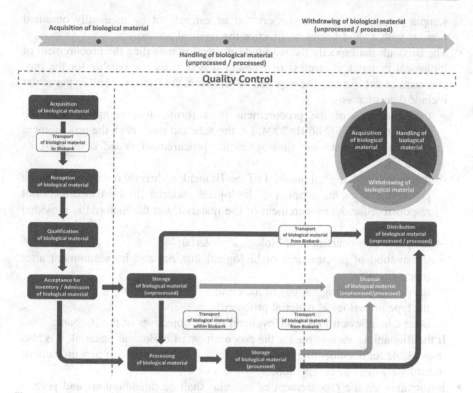

Fig. 10.1 The example of the general life cycle of biological material (BM) and exemplary critical processes

10.1.1 Acquisition of the Biological Material and Related Data

- If the Biobank is responsible for procurment/collection of biological material, it shall establish, implement, and use the procurement procedures for each biological material procured. The above-mentioned procedures should:
  - be established by appropriate and qualified employees of the Biobank, taking into account the requirements of subsequent users (if known);
  - take into account the requirements arising from the characteristics of biological material and the required quality and quantity of the material, taking into account its final destination, if it has been pre-defined.
- The procurement of biological material shall only be carried out by qualified personnel (see Chap. 4 MBQM). The procurement of biological material by non-Biobank employees is referred to in Sect. 5.4 MBQM.
- In the case of human biological material that requires clinical assessment or diagnosis (e.g., tissue sections when it is necessary to determine the margins of the lesion), the procurement of biological material, pathology assessment, and

sample preparation shall be supervised or carried out by medically qualified persons with a valid license to practice the medical profession.

- The Biobank shall specify the required information regarding the procurement of biological material, regardless of whether it is directly responsible for the procurement or not. Among the required information, the Biobank shall in particular include data such as:
  - the timestamp of the procurement in a format that complies with the requirements of ISO 8601: 2004, i.e. the date and time when the procurement began and the date and time when the procurement ended or at least its duration;
  - the donor of biological material (if the Biobank is directly responsible for the procurement) or the supplier of biological material (if the Biobank is not responsible for the procurement of the material, but the material is provided to it);
  - the place of procurement of biological material;
  - the method of procurement of biological material and its securement after procurement;
  - the person procuring biological material;
  - the type of biological material procured;
  - other data relevant to the achievement of the objectives of the Biobank.
- If the Biobank is responsible for the procurement of biological material, it is also responsible for documenting the above information at the time of procurement or immediately after its completion.
- Information on the procurement of material shall be unambiguous and permanently associated with it.
- Information about the procurement process is subject to verification after biological material is delivered to the Biobank and during the qualification constitutes the basis for the decision to accept or reject the material delivered to and received at the Biobank (see Chap. 8 MBQM).

7.2.1.1, 7.2.1.2, 7.2.3.1, 7.2.3.3, annex A.2 and annex B.2 ISO 20387:2018

The Most Common Practices
- The required data on the donor most often include:
  - personal data/ID—in the case of prior anonymization of the donor;
  - data on the donor's consent to the procurement and use of biological material and associated data for a specific purpose;
  - donor's biometrics data such as age, gender, height;
  - medical data of the donor (medical/health questionnaire), and if the donor is a patient, also data on the diagnosis and current course of the disease, conducted treatment and its results;
  - donor's epidemiological data (socio-environmental interview), including data on possible exposures, smoking, diet.
- The required data on the procured biological material most often include data on:

- the type of the material;
- infectivity status of the material;
- the unique identifier of the material (if it has been previously assigned).
- The required data for the procurement of biological material most often include data on:
  - the person identifying the donor, data on the person procuring and protecting the material after procurement, and data about the possible supervisor;
  - the date, time, and duration of the procurement;
  - the method of material procurement and its securement after procurement and storage conditions before delivery to the Biobank (i.e., before and during its transport);
  - the tools and equipment used during the procurement and protection of the material;
  - products and materials which are in direct contact with the procured material as well as on the container in direct contact with the procured material.
- For example, the biological material procurement report usually contains the following information (see also Most common practices in subchapter 8.2 MBQM):
  - procurement report number (assigned by the Biobank);
  - data on the research project, in particular its ID and name;
  - designation and the number of the version/edition/issue of the procedure/instruction according to which biological material was procured;
  - identification data of the person who verified the identification of the donor: name, position, signature of that person;
  - data on the place/location of the procurement:
    - (a) identification data of the medical entity where the procurement took place: name and contact details;
    - (b) the name of the organizational unit of the medical entity (e.g., procurement point, treatment room, endoscopic room, operating theater) and its contact details if different from the data of the medical entity.
  - data on procurement:
    - (a) identification data of the person procuring biological material: name, job title, position;
    - (b) information about the procuring person's knowledge and compliance with the applicable version/edition/issue of the procedure/instruction pertaining to the procurement and protection the procured biological material, the person's signature;
    - (c) date and time of procuring biological material (in a format compliant with the ISO 8601: 2004 standard) or, in the case of longer procurement, date and time of beginning and ending the procurement of biological material (in a format compliant with the ISO 8601: 2004 standard);
    - (d) information on the course of the procurement, including the occurrence/absence of complications/side effects during the procurement.
  - data on the procured biological material and related data:

(a) biological material identification data:
   i. the type of material procured (e.g., whole venous blood, skin slice, mucosa slice, bronchoalveolar lavage);
   ii. the volume of material procured: mass in grams (if applicable), the volume in milliliters (if applicable);
   iii. the number of containers with procured material (the number of pieces);
   iv. the method/methods of securing the material after procurement (e.g., smearing on a slide with subsequent fixation, placing the material on a transport medium, placing in a fixative [e.g., formaldehyde, glutaral-dehyde solution], cooling to a temperature of + 2 °C to + 8 °C, freezing, providing protection against light).

- identification data of products and materials in direct contact with the procured biological material, e.g. container data in which biological material is placed, fixative agent data, transport medium data;
- identification data of the person responsible for securing the procured material with associated data until they are transferred to the Biobank: name, position, and signature of that person.

• The procurement report is often combined into one whole with the transport report to the Biobank of the protected biological material.

10.1.2 Transport of Biological Materials and Associated Data

10.1.2.1 Transport from and to the Biobank (Shipping)

• The Biobank shall establish, implement, and use procedures regarding the transport process to and from the Biobank for each biological material and associated data.
• These procedures shall take into account the properties of materials and the type of transport (air, water, land transport). The procedures shall describe how to handle a delivery at every stage, i.e. when preparing it for shipment, packaging, sending, and receiving.
• The procedures concerning transport shall be carried out by trained personnel (see Sect. 4.6 MBQM).
• The Biobank shall determine the transport conditions (e.g., its maximum time, acceptable temperature range, acceptable humidity range, necessity/lack of protection against light), bearing in mind the need to ensure maintenance of integrity of biological material.
• Each transport event shall be tracked and monitored in relation to the parameters relevant to the integrity of a specific biological material, such as, for example, time, temperature, humidity, or access of light.
• The documented temperature verification shall take place at least at the point of departure and destination of transport. Temperature monitoring during transport shall at least include documenting the minimum and maximum temperature.
• Any deviations from the value ranges defined for individual parameters are recorded and documented (see Chap. 11 MBQM).

- Before submitting biological material and related data from the Biobank, the Biobank shall determine the conditions for their distribution and receiving with the address/recipient of the material and data (e.g., taking into account transport conditions and the type of accompanying documentation).
- The Biobank shall keep records of the control of the transport process of each biological material from the moment it is sent to the time the package is received.

7.4.1, 7.4.2, 7.4.3, 7.4.5, 7.4.6 ISO 20387:2018

The Most Common Practices
- In the context of external and internal transport of biological material, it is required to document:
  - transport method/shipment specification, including:
    (a) UN biological threat code (e.g., UN 3373—diagnostic preparations, clinical preparations, or substances of biological origin);
    (b) packing instructions (e.g., instruction PI650—United Nations Packing Instruction 650);
    (c) prohibitions (e.g., no irradiation);
  - temperature during transport, including maximum and minimum value;
  - temperature on delivery, including maximum and minimum value;
  - timestamp—dates and times of the starting point and end of transport (in a standard format compliant with ISO 8601:2004 standard);
  - special requirements (e.g., humidity, light access, maximum shipping time).
- Due to the large number of necessary steps to be taken, checklists are a useful element related to transport procedures; checklists are forms/prints containing lists necessary to perform activities during a specific stage with the possibility of marking their completion.
- The most commonly used method of temperature monitoring during transport is using data loggers. Another possible way are portable temperature recorders (most often with the readings making it possible to obtain the information concerning the current temperature as well as minimum and maximum temperatures, both outside and inside the container) with sensors placed inside the transport container. It is necessary to ensure supervision of data loggers and temperature recorders, including systematic and documented verification of their battery condition and their calibration.
- It is important to remember to adapt the way the material and data are packed to the requirements of carriers, especially in the case of transport by air.

10.1.2.2 Transport Within the Biobank
- The Biobank shall establish, implement, and use procedures for transport within the Biobank (procedure for handling/transfer/internal transport) for each biological material and associated data.
- In order to ensure the integrity of biological material being transferred, any transfer of biological material shall be carried out with strictly defined parameters.

- For each biological material, the Biobank retains documented information (records) of control of the internal transport process from its commencement to its completion. Documented information (records) of internal transport control may include all or selected elements listed for the external transport process.
- Any deviation from the specified parameters is treated as an occurrence of noncompliance (see Chap. 11 MBQM). Biological material shall not be left unattended outside the designated places indicated in relevant procedures.

7.4.2, 7.4.4 ISO 20387:2018

10.1.3 Reception, Qualification, and Acceptance of Biological Material and Associated Data for Inventory

- The Biobank shall establish, implement, and use procedures for the reception of biological material and related data delivered to the Biobank, qualification of the received material and data as well as admission of qualified material and data to the Biobank.
- Above-mentioned procedures shall take into account the origin and ownership status of biological material and associated data (e.g., material procured by the Biobank or under its responsibility, material transferred to the Biobank under a cooperation agreement, material entrusted to the supervision of the Biobank; see Sect. 5.4 MBQM).
- After the reception of biological material and associated data delivered to the Biobank, the Biobank performs qualification of the received material and data by verifying whether they meet specific acceptance criteria.
- The Biobank shall define itself or with the participation of stakeholders, the acceptance criteria for the admission of biological material and associated data, taking into account the issue of biosecurity and property rights.
- Until the confirmation of the fulfillment by the delivered and received biological material and associated data of legal, ethical, quality, and documentation requirements, the Biobank shall store them in a manner ensuring its biological integrity in a separate place outside the place of routine storage of biological material and associated data.
- In the case of a positive result of the qualification of the received material and associated data resulting in their acceptance for the Biobank's inventory, the Biobank registers the accepted material and data and assigns them a unique identification (see Chap. 8 MBQM).

7.3.2.1, 7.3.2.2, 7.3.2.3, 7.3.2.4, 7.3.2.5, 7.3.2.6 ISO 20387:2018

The Most Common Practices
- The report on the reception of biological material and associated data by the Biobank is often combined into one whole with the report on the qualification of

the received material and data as well as with the report on the admission of positively qualified material and data.

- During the reception and qualification of biological material and associated data delivered to the Biobank, the verification is subject to, among others:
  - the condition and labeling of the transport container;
  - the condition and labeling of the containers in direct contact with biological material;
  - the compliance between the content of the transport container and the information in the documentation attached to the shipment, as well as the information in the order (if applicable).
- In the established and implemented *Report on the reception, qualification, and admission of biological material and associated data*, documenting practices include:
  - data of the person receiving the material and data in the Biobank;
  - the date and time (in a format compliant with ISO 8601: 2004) for material and data reception;
  - the verification performed during the reception process regarding:
    - (a) the length and conditions of material and data transport;
    - (b) the external condition of the transport packaging;
    - (c) the presence of a transport label and the correctness (compliance, readability, and completeness) of the data contained therein;
  - actions taken in the event of deviations/inconsistencies found during reception of material and data;
  - data of the person conducting the qualification of the material and data received in the Biobank;
  - the date and time (in a format compliant with ISO 8601: 2004) of the commencement of material and data qualification;
  - the verification performed during material and data qualification regarding:
    - (a) the condition of the container with the material;
    - (b) the presence of the identification label on the container with the material as well as the correctness (compliance, readability, and completeness) of the data contained therein;
    - (c) the type and condition of the material received;
    - (d) the presence and correctness (compliance, readability, and completeness) of the data associated with the material;
  - the result of the qualification of the material and data, i.e. acceptance, conditional acceptance or the refusal of accepting (rejection) of the material and data received;
  - actions taken in the case of deviations/noncompliance during material and data qualification;
  - the date and time (in a format compliant with ISO 8601: 2004) for the acceptance for the Biobank's inventory of qualified material and data or their rejection;
  - the unique identification of the material and data accepted for inventory.

Frequently Asked Questions
- **May the Biobank refuse to accept biological material and associated data, and, if yes, when?**
  The Biobank may refuse to accept the material and data. Such activity has to be included in the Biobank's internal procedures and the external entities supplying the material and data have to be informed in advance about such possibility, e.g. such possibility is included in the written information for the material procurement entity or in the cooperation agreement with such entity. Both above-mentioned procedures as well as information for the entity (contained in the contract with the entity) should clearly specify the situations in which the Biobank refuses to accept the material and the further course of action in such a case.
- **What could be the reasons for refusing to accept biological material and related data at the Biobank?**
  The reason for refusing to accept biological material and associated data at the Biobank may be, for example, an irreversible damage to the transport packaging or incorrect transport conditions that result in an irreversible damage to biological material.

10.1.4 The Process of Handling of Biological Material and Associated Data

10.1.4.1 Processing of Material and Data
- The Biobank shall establish, implement, and use procedures for biological material processing. The above-mentioned procedures shall take into account evidence-based processing methods that have been published in the scientific or professional literature as well as in the published standards. The above-mentioned methods shall also take into account the requirements specified in the cooperation agreement with the supplier/recipient/user, as well as the requirements contained in the manufacturer instructions of the reagent and other materials.
- The Biobank shall monitor the critical activities of processing biological material and document the performance of all activities and all relevant parameters. Each stage of processing shall be documented individually.
- The Biobank shall document in a standard format in accordance with ISO 8601: 2004, the date of every stage of processing of each biological material and the duration of individual stages.

The Most Common Practices
- The standards to which the Biobank may refer when choosing or establishing its own method of processing biological material include, among others, standards such as:
  - ISO 20166-1: 2018 Molecular in vitro diagnostic examinations—Specifications for pre-examination processes for formalin-fixed and paraffin-embedded (FFPE) tissue—Part 1: Isolated RNA;

- ISO 20166-2: 2018 Molecular in vitro diagnostic examinations—Specifications for pre-examinations processes for formalin-fixed and paraffin-embedded (FFPE) tissue—Part 2: Isolated proteins;
- ISO 20166-3: 2018 Molecular in vitro diagnostic examinations—Specifications for pre-examination processes for formalin-fixed and paraffin-embedded (FFPE) tissue—Part 3: Isolated DNA;
- ISO 20186-1: 2018 Molecular in vitro diagnostic examinations—Specifications for pre-examination processes for venous whole blood—Part 1: Isolated cellular RNA;
- ISO 20186-2: 2018 Molecular in vitro diagnostic examinations—Specifications for pre-examination processes for venous whole blood—Part 2: Isolated genomic DNA.

10.1.4.2 Storage of Material and Data

- The Biobank shall establish, implement, and use procedures for storage and location tracking of biological material and associated data. These procedures should include at least:
  - information on the labeling of biological material and associated data containing at least its unique identifier;
  - the type of the container and storage conditions of biological material;
  - the mechanisms for ensuring traceability;
  - the plan to maintain the required storage conditions in the event of a short-term emergency.
- To avoid losing biological material and associated data, the Biobank or a legal entity that the Biobank is a part of shall establish and implement a disaster protection plan using alternative protection methods.
- Storage locations and processes of biological material must be designed to ensure the maintenance of its biological integrity and to minimize the risk of cross-contamination.
- The Biobank shall measure, monitor, and document critical parameters related to the storage of biological material, e.g. length, temperature, humidity. The Biobank shall also document the dates and time of critical activities (in accordance with the format included in the ISO 8601:2004 standard) and the data of persons performing them.
- The Biobank shall document and verify the storage location of each biological material and related data.
- The Biobank shall ensure traceability of all biological material and related data throughout the entire storage period.
- The Biobank shall establish, implement, and use procedures to withdraw the donor/patient consent to storage and the use of biological material and related data.

7.6.1, 7.6.2, 7.6.3, 7.7.1, 7.7.2, 7.7.3, 7.7.4, 7.7.5, 7.7.6, 7.7.7, 7.7.8, Annex A.4
ISO 20387:2018
8.5.1 ISO 9001:2015

The Most Common Practices
- The Biobank should define and implement a schedule for conducting an inventory of biological material and accociated data stored and then carry out such an inventory in accordance with the prepared and implemented schedule and document the whole process as well as the results.
- The process of placing biological material in the storage site and withdrawing it therefrom is usually automated and carried out using robots.

10.1.5 Transport of Material and Data Within the Biobank

- The transport of biological material and associated data within the Biobank is discussed in Sect. 10.1.2.2 MBQM.

10.1.6 Process of Withdrawing Biological Material and Associated Data

- The Biobank shall establish, implement, and use procedures for distribution of biological material and associated data, ensuring that only established (agreed) biological materials and related data are distributed in accordance with the terms of the contract concluded with the recipient/user.
- The transport of biological material and associated data from the Biobank is discussed in Sect. 10.1.2.2 MBQM.
- The Biobank shall establish, implement, and use procedures for the disposal of biological material and associated data.
- The Biobank shall also ensure the security of associated data transfers (see Chap. 15 MBQM).

4.3.2, 4.3.3, 7.3.3.3 ISO 20387:2018
8.2.3.1, 8.5.1 h), 8.6 ISO 9001:2015

10.2   Quality Control

- In order to demonstrate the suitability of biological material and associated data for the intended purpose, the Biobank shall establish, implement, and use procedures defining quality control activities (QC) in the area of all critical biobanking processes.
- The Biobank shall include QC criteria in the above-mentioned procedures.
- The Biobank shall establish and implement a schedule for carrying out QC activities related to processes and perform them in accordance with the prepared and implemented schedule.

- The Biobank shall keep documented information regarding activities and results of QC and ensure that information regarding the results of QC is provided to the recipient/users, if it is included in the cooperation agreement.
- In a situation when the criteria included in the implemented procedure have not been met, the Biobank shall take an action to ensure adequate control over distribution of noncompliant biological material and associated data or overreporting of incorrect data. The Biobank shall ensure that limitations found during QC have been clearly documented and forwarded to the recipient/user, if it is provided for in the cooperation agreement.
- The Biobank shall also ensure that the responsibility for accepting biological material and associated data transferred from the Biobank, along with the documented and forwarded restrictions, rests with the recipient/user during the process of distribution of biological material and related data.
- The Biobank shall periodically analyze the results of QC in terms of trends and those used as output data in the process of continuous improvement.
- Quality control of data associated with biological material shall focus on the accuracy, completeness, and consistency of these data.

7.1, 7.8.1, 7.8.2, 7.8.3 ISO 20387:2018
8.5.1 b), 8.5.1 c) ISO 9001:2015

The Most Common Practices
- Routine quality control of the process of collection, transport, and centrifugation of biological material in the form of venous whole blood samples can be performed through the periodical (e.g., during a management review) determination of the percentage of samples with hemolysis and the subsequent trend analysis. If, during the subsequent assessments, the percentage of samples with hemolysis increases, it should be assumed that at one of the stages (collection, transport, centrifugation), some unfavorable changes occurred, affecting the quality of biological material and it is necessary to implement corrective and preventive measures.
- Usually, quality control for a newly introduced critical technological process (e.g., a new method of DNA isolation) is carried out at least every 3 months. After 6 months from the introduction of a new critical technological process, if the trend analysis does not show significant deviations/inconsistencies (e.g., the quantity and purity of the DNA isolated by the new method meets the acceptance criteria), the frequency of performing quality control can be extended to 6 months and finally with the condition of no deviations/inconsistencies in the analysis of trends—up to 12 months.
- At the same time, quality control every 12 months is recommended for processes in which the Biobank has at least one and a half year of experience.
- If the trend analysis indicates that there are significant deviations/inconsistencies in the process, it is recommended to increase the frequency of performing QC.
- It is recommended for the Biobank, as part of the quality control system, to possess appropriate quality control materials (e.g., internal control materials).

The Biobank should periodically check the above-mentioned materials to assess the significant quality characteristics of biological material, including stability, the performance of the processing method, and the accuracy/precision of quality control procedures.

- The participation of the Biobank in external quality assessment (EQA) programs, external proficiency testing programs, or inter-laboratory comparisons can ensure that the Biobank obtains objective evidence of the comparability of biological material quality.
- The Biobank can also develop its own approaches to demonstrate assurance and maintain comparability of biological material quality. The above-mentioned approaches may be based on the use of available certified reference materials or control materials that are regularly tested in EQA or on the use of previously tested own samples or samples shared with other Biobanks.

Frequently Asked Questions

- **Which incorrect quality control results must be forwarded to the recipient?**
  All results that do not meet/are out of specification should be reported. These may be, e.g., storage temperature, cell viability, or the number of cells that do not comply with the specification, as well as the presence of hemolysis or deviation in the concentration and purity of isolated DNA.

References

ISO 20387:2018 Biotechnology—Biobanking—General Requirements for Biobanking.
ISO 9001:2015 Quality management systems—Requirements.

Deviations, Nonconforming Product/Data or Service

11

Abstract

Nonconformities and deviations are typical issues of any organization, but they are also a starting point for continuous improvement. This chapter MBQM Deviations, nonconforming product/data or service discusses the issues related to the identification of any deviations and nonconformities in the Biobank as well as indicates how to deal with them. It contains information on the implementation of correction and corrective actions, as well as on the submission, processing, and handling of complaints. The process of identifying and rejecting nonconforming products/services and accompanying documentation was also descried in detail in the prepared study. The chapter presents numerous practical examples, guidelines for action, and forms, such as a nonconformance card template, which the Biobank can implement and use in its activities.

11.1 Deviations/Nonconformities

- The Biobank shall develop, implement, and apply a procedure describing the measures to be taken in the event of identifying deviations/nonconformities, their documentation (including registration) and taking corrective actions.
- The procedure shall apply to every process taking place at the Biobank.
- The procedure shall specify the persons responsible for taking corrective and correction actions, their monitoring and evaluation of their effectiveness, as well as the method of documenting individual activities.
- The Biobank shall counteract the occurrence of deviations/nonconformities by establishing the proper control of each process taking place at the Biobank.

4.1.1, 4.1.7, 5.9 c), 6.2.3.1, 7.11.1.6, 8.7 ISO 20387:2018
8.7.1, 10.2 ISO 9001:2015

© Wroclaw Medical University 2023
A. Matera-Witkiewicz et al., *Manual of Biobank Quality Management*,
https://doi.org/10.1007/978-3-031-12559-1_11

Table 11.1 The exemplary rules pertaining to dealing with deviations/nonconformities in the Biobank

Person responsible	Stages	Status
Employee, colleague, supplier	Deviations/nonconformities report	Transfer to the Biobank Manager
The Biobank Manager or a designated person	Determining reasons for deviations/nonconformities	Accepting the application
The Biobank Manager or a designated person	Planning correction and corrective actions and appointing a person to implement the above actions	Implementation of planned activities
Person implementing correction and corrective action	Summary of correction and corrective actions carried out, evaluation of their effectiveness	Implementation of planned activities
Person implementing the above activities	Completion of works or launching a project of improvement actions	Closed application—an improvement project was initiated or Closed application—an improvement project was not launched
The Biobank Manager	Final acceptance of all actions taken	Approved and closed report

The Most Common Practices
- The exemplary rules pertaining to dealing with deviations/nonconformities in the Biobank: the responsible person was indicated, the stages of dealing with deviations/nonconformities and the status of each stage were described (Table 11.1).

11.2   Corrective and Correction Actions

- After finding deviations/nonconformities, the Biobank shall immediately take corrective and correction actions.
- The Biobank shall reanalyze the risk of identified critical nonconformity (see Sect. 2.4 MBQM).
- The effectiveness of corrective actions taken shall be assessed during the management review.
- Monitoring the effectiveness of corrective actions in relation to critical nonconformities shall take place over a specified period of time and frequently enough.

7.11.2.1, 8.7.1, 8.7.2, 8.7.3, 8.9.2 ISO 20387:2018
8.7.1 a), 10.2.1 ISO 9001:2015

The Most Common Practices
- One of the practices used in Biobanks is the monitoring of corrective actions, which takes place for a specific period of time, determined by the Biobank itself depending on the significance of nonconformity. After this time, the team summarizes the effectiveness of the measures applied by documenting this fact in the nonconformity card. If the summary result is positive, the nonconformity card is closed. If the result is negative, additional corrective actions should be taken in order to effectively prevent deviation/nonconformity.
- The process of dealing with deviation/noncompliance (corrective actions) may end with undertaking improvement actions (see Chap. 13 MBQM).

Frequently Asked Questions
- **What form of verification can be used to monitor the effectiveness of corrective actions?**
  Risk analysis is useful to monitor the effectiveness of corrective actions. It may also influence the decision to terminate the monitoring process.

11.3 Complaints

- The Biobank shall provide the possibility to submit a complaint (see Sect. 1.4 MBQM).
- The Biobank shall specify the rules for dealing with complaints, including the rules for receipt, examination, and decision regarding the complaint.
- The method of dealing with the complaint shall be documented.

   7.11.1.4, 7.13, 8.2 ISO 20387:2018
   8.2.1 c), 10.2.2 ISO 9001:2015

The Most Common Practices
- An example of how to deal with a complaint may be a situation in which the Biobank describes the principles of managing complaints in the internal procedure of QMS. In such an event, the Biobank refers to the Complaint form. This form is always attached to the documentation accompanying the biological material issued.
- The form may also be available on the Biobank website.
- The Biobank enables the submission of complaints in oral, paper, and/or electronic form, using the form available on the website. Each of these complaints is recorded in the internal register of the Biobank. During registration, the complaint is assigned to a specific category (e.g., donation of biological material, collection of biological material, completeness of documentation regarding the collected biological material, transport of biological material, release of biological material to an external entity.
- The Biobank shall keep a register of complaints. The register should include: the subject of the complaint, a brief description of the complaint, details of the

department/person responsible for the complaint, information whether the complaint has been closed.

The Frequently Asked Questions
- **Is it possible for the entity submitting a complaint to have access to the documentation regarding the complaint procedure?**
  Yes, the entity submitting a complaint has the right to inspect the complaint procedure at every stage, on the spot in the Biobank.

11.4 Nonconforming Output/Service

- The Biobank shall establish and implement a nonconforming/incompatible service procedure.
- The procedure shall define acceptance criteria for a nonconforming product/inconsistent service, i.e. indicate the possible deviations with which the noncompliant product/noncompliant service will be authorized (accepted) for use, and for which it will be rejected.
- The procedure shall specify who is responsible for making a decision about the acceptance or rejection and how to document the handling of a nonconforming product/service that is not compatible.

4.1.1, 7.11.1.1, 7.11.1.5, 7.11.1.6 ISO 20387:2018
8.7.1, 8.7.2 ISO 9001:2015

11.4.1 Identification of Nonconforming Products/Services and Accompanying Documentation

- Identification of a nonconforming product/service can be made by each of the Biobank employees.
- After finding the deviations/nonconformities, a Biobank employee shall report the fact of identifying a nonconforming product/service to the Biobank Manager in the form of a document.
- The Biobank Manager (the analysis is allowed in a wider group, e.g., with the department responsible for a given process, with the department responsible for quality) analyzes the application concerning a nonconforming product/ service.
- Based on the analysis, referring to the approval criteria set in the procedure, the Manager decides whether to accept or reject the application as unjustified.

7.11.1.5, ISO 20387:2018
8.7.1 ISO 9001:2015

The Most Common Practices

- If the declaration accepting the nonconformity as affecting the quality and safety of biological material and the documentation or quality of the service provided by the Biobank is accepted, a risk analysis should be carried out and a record should be created informing whether the nonconforming product/service may be admitted (see Sect. 2.4 MBQM).

The Frequently Asked Questions

- **What risk analysis methods can be used to evaluate a nonconforming product/service?**
  When analyzing the risk for a nonconforming product/service, methods such as FMEA, 5 Why, Ishikawa diagram may be used. The method of analysis chosen should enable the performance of the most accurate assessment in order to decide whether to accept or reject a nonconforming product/service.
- **What are the approval criteria for nonconforming outputs?**
  The approval criteria for nonconforming outputs are the sets of information which help the Biobank Manager to make a decision about the nonconformity. For example, the range of cells viability should be between 85 and 100% but if biological material is very valuable, the Biobank Manager could enter the information into the procedure to accept lower values.

11.4.2 Rejection of Nonconforming Product/Service

- The Biobank shall determine the rules pertaining to dealing with a nonconforming product/service after rejection.
- Rejection of a nonconforming product/service shall be documented and approved by the Manager of the Biobank.
- An incompatible product shall be separated from other authorized products and clearly marked to avoid uncontrolled use (e.g., labeled "disqualified").
- A nonconforming product that has been disqualified from further use (e.g., it cannot be used for scientific research) shall be disposed of, except in the case where the Biobank determines its suitability for other purposes (e.g., intended as a quality control material).

7.11.1.1, 7.11.1.2, 7.11.1.5 ISO 20387:2018
8.7.1, 8.7.2 c) ISO 9001:2015

The Most Common Practices

- An example of a nonconforming product accepted for release may be partially damaged biological material. Damage can be mechanical (e.g. material is fragmented, damaged during transport) or qualitative (e.g., deterioration of material characteristics resulting from storage at the wrong temperature, damage during processing). The basis for the acceptance of a noncompliant product for release is its scientific value, which exceeds the inconvenience associated with damage to the material.

References

ISO 20387:2018 Biotechnology—Biobanking—General Requirements for Biobanking.
ISO 9001:2015 Quality management systems—Requirements.

Audits

12

Abstract

A systemic approach to activities that are aimed at guaranteeing the reproducible quality of processes and biological samples in the Biobank can be developed by meeting the requirements set out in ISO 9001, 20387, and 19011. One of these requirements is to carry out quality audits. Internal audits allow the organization to achieve its strategic and operational objectives and to ensure its continuous improvement. Their aim is to define the causes, to prevent the occurrence of adverse events or nonconformities, and to determine the supposed process risk. The chapter Audits points out to the most important aspects of internal audits in the Biobank and discusses all key stages and best practices, including the process of creating an audit program, preparation for audits, performance and documentation of audit procedures, as well as post-audit activities. In addition, this section describes the competence and qualifications of auditors.

- The Biobank shall be subject to regular audits, covering the entire QMS documentation, which regulates the functioning of the Biobank and processes, including technological ones.

8.8.1, 8.8.2 ISO 20387:2018
9.2 ISO 9001:2015

The Most Common Practices
- At the Biobank, the following types of audits are distinguished:
  - first party audit (see Glossary)—an internal audit carried out by its own team of internal auditors in order to confirm the effectiveness of QMS operation and the functioning of other processes in the Biobank;

- second party audit (see Glossary)—an audit carried out by the Biobank on suppliers/subcontractors of materials or services in order to verify the areas of the supplier's operation;
- third-party audit (see Glossary)—an audit confirming the compliance of the implemented management system with the requirements of a selected standard, conducted by an independent organization (certification institution).

12.1 Internal Audit

- Within the scope of internal audits, the Biobank shall plan, implement, and maintain an Audit Procedure (the so-called Audit Program) containing information such as:
  - audit frequency:
    - (a) the Biobank shall plan audits and their frequency adequately to the status and importance of processes to be audited;
    - (b) it is recommended that the Biobank shall plan, implement, and maintain an Audit Schedule;
  - audit methods to be used;
  - the role and responsibility of the person in charge of the audit program;
  - audit planning:
    - (a) the Biobank shall prepare an audit plan containing the following information:
      audit objectives;
      the scope of audit;
      the name and position of persons directly responsible for areas/processes related to the objectives and scope of an audit;
      the name and surname of the members of audit team;
      the date of an audit;
      the name of the unit covered by an audit;
      if possible, the date of the commencement and the duration of each activity during an audit;
      confidentiality requirements;
  - the audit plan shall be disclosed to the auditee;
  - *reporting:*
    - (a) the Biobank shall prepare a report on each audit carried out;
    - (b) the report shall be prepared by an auditor or a team of auditors;
    - (c) the report shall contain the auditors' observations and observations as well as the identified nonconformities;
    - (d) the report shall be prepared in a timely manner (if there is a delay, the reasons should be forwarded to the appropriate person);
    - (e) the audit report shall contain the following information:
      audit objectives;
      audit scope;
      list of audited (name and surname, position/function);

list of auditors (names, functions: lead, auxiliary, technical expert);
date and place of audit;
audit criteria;
observations (description and evidence);
audit conclusions;
nonconformities/recommendations (description and evidence);
signature of the auditor/auditors.

The audit report shall be forwarded to recipients in accordance with the audit procedure or audit plan

- The Biobank shall specify the criteria and scope for each audit.
  - the Biobank shall determine the scope and detail of an audit. It includes factors such as: area/areas; process/processes; persons. The scope of an audit shall be consistent with the audit program. The scope of an audit shall be integrated with its objectives.
  - the Biobank shall specify audit criteria, i.e. references against which compliance is determined (e.g., procedures, guidelines, standards, norms, codes).
- The Biobank shall ensure that the results of audits are reported to the appropriate management.
- The Biobank shall implement corrective actions related to the nonconformities identified during the audit.
- The Biobank shall establish, implement, and apply a procedure for defining the rules aimed at identifying nonconformities, supervising and running corrective, correction, and preventive actions which eliminate the causes of nonconformities (see Chap. 11 MBQM).
- The auditor shall document the identified nonconformities in the best manner.
- The auditor shall refer and classify the detected nonconformities to the requirements of norms/standards/procedures/instructions.
- The Biobank shall keep the records of all events/nonconformities on the introduced form used for recording the nonconformities events and corrective actions taken.
- The Biobank shall keep the register specifying the nonconformities events and describing the effectiveness of corrective actions taken.
- The Biobank shall maintain the records regarding the audit process.

6.4.1.6 b), 8.8.1, 8.8.2 ISO 20387:2018
9.2.2, 10.2.2 ISO 9001:2015
5.1, 5.2, 5.3, 5.4, 5.5, 6.3.2.2, 6.3.3, 6.5.1, 6.5.2, 6.5.3 ISO 19011:2018

The Most Common Practices
- It is recommended to introduce a form including the schedule of audits constituting the basis for audits (containing information such as: the year for which the schedule is established, the area being audited, the scope of an audit, the planned date of an audit, the name of the auditor appointed).

- Audits can be carried out quarterly, every 6 months, or once a year or in response to: an inconsistent incident, any change or deviation from the procedure; changes in the law, e.g., in terms of ethics, health, and safety.
- An internal audit is carried out in accordance with the schedule of internal audits, a planned audit takes place at least once a year and, if necessary, there are ad hoc audits held as well.
- The frequency of audits depends on: the relevance of the area/process; the results of previous audits; the implementation of new methods/processes/technologies; organizational changes in the Biobank; the implementation of corrective actions nonconformities identified or problems found in the pre-defined area.
- The scope of the audit plan will depend on the size and specificity of the controlled area in the Biobank, as well as the nature, functionality, complexity, and level of advancement of QMS to be audited.
- The purpose of the audit report is to make it possible for the Biobank to fully use all the information provided by the report, and enable its Management to correctly assess the processes in audited areas as well as to take appropriate action.
- The audit report is prepared within no more than 21 days of the inspection. If it is delayed, it is recommended to communicate the reasons for the delay to the auditee.
- The results of audits are collected by the person or team responsible for QMS in the Biobank and stored in accordance with sec. 3 of MBQM.
- Audit reports can be accessed by the Biobank Manager and the person/team responsible for QMS in the Biobank.
- It is recommended that the nonconformity card contains the following information:
  - Section I—Completed by the employee reporting inconsistency or receiving an audit report: the date of notification; nonconformity card number (recommended: sequence number/year); the description of nonconformities; reporting nonconformities (external audit, internal audit, submitting employee/colleague/external person, name and surname of the reporting person);
  - Section II—Completed by the Biobank Manager: the acceptance of nonconformities (date and signature); the description of the cause of nonconformities; the activities proposed; expected date of the removal of nonconformities;
  - Section III—Completed by the employee responsible for reaction/removal of nonconformities: the description of actions taken; information whether the actions taken had a positive effect; date and signature of the employee responsible for reaction/removal of nonconformities;
  - Section IV—Completed by the Biobank Manager: final acceptance of all actions taken; approved report/closed application (in the case of non-acceptance, the reason needs to be presented); the date of closing the notification/removing nonconformities; date and signature of the Biobank Manager.

- It is recommended that a nonconformities register be introduced, containing the following information: nonconformity card number; the source of nonconformity; the report number (if applicable); the person responsible for the implementation of actions; the description of the discrepancy; information whether the corrective actions were taken (YES/NO); information whether it was an adverse event or reaction (YES/NO); the date of closing the nonconformity card; comments.

Frequently Asked Questions
- **Can the scope of the audit be changed?**
  Yes, the scope may change at the latest during the audit opening meeting.
- **Which areas in particular require regular auditing?**
  Areas that have a significant/critical impact on the quality and safety of biological material and associated data.
- **How to proceed when the auditee has comments on the presented audit plan?**
  It is recommended that all remarks and corrections of the auditee to the audit plan be settled between the lead auditor and the auditee at the opening meeting at the latest.
- **Does the audit plan have to be submitted on a special form?**
  No, if the Biobank can document that the audit plan is prepared and communicated to the auditee, e.g. in the form of an email.
- **Should auditors make copies of audited documentation?**
  Auditors do not have to make copies of the documentation checked. The report only needs a detailed description of the record being checked.
- **Should the auditor read the previous report before the next audit?**
  Before undertaking a scheduled audit, the internal auditor should read the report from the previous audit and start checking the area by verifying the removal of nonconformities (if demonstrated) from the previous audit.
- **What does it mean that the auditor needs to document the identified nonconformity?**
  The auditor should be able to answer the following questions: What has been identified? Where has it been identified? On what basis? What is the evidence for this?

12.2 Auditor's Competences and Qualifications

- The Biobank shall determine the competences and qualifications of auditors
- The auditor shall have documented qualifications and have the knowledge and skills necessary to achieve the intended audit results (acquired during education, professional experience, training for auditors, and experience in auditing)

- The auditor shall improve his competences through continuous professional development
- The auditor shall have appropriate personality traits and apply knowledge and skills in practice
- The lead auditor shall have additional knowledge and skills to manage the team of auditors
- The competences of each of the Biobank auditors shall be subject to a regular assessment, whereby:
  - the assessment shall be documented;
  - assessment criteria shall be qualitative and quantitative;
  - the assessment shall be made using one or more methods.
- Audits shall be carried out by trained and competent staff who are independent of the area subject to audit
- Auditors are responsible for the following:
  - conducting an audit in accordance with the audit plan and the instructions of the lead auditor
  - the documentation of nonconformities
  - maintaining confidentiality of information obtained during auditing activities
  - preparing an audit report.

9.2.2 c) ISO 9001:2015
7.1, 7.2, 7.3, 7.4, 7.5, 7.6 ISO 19011:2018

The Most Common Practices
- It is recommended to describe the competences of auditors on an *Auditor's competences* form. Competences of the auditor are determined by the person/ team responsible for QMS in the Biobank.
- The assessment of auditors' qualifications is made in the form of observations during internal audits. The assessment is recorded on a dedicated form. The assessment of internal auditors' qualifications is performed by the person/team responsible for QMS in the Biobank or an internal auditor who has met the requirements and obtained the right to assess another auditor.
- The division of areas takes into account the independence and competence of auditors and the efficient use of resources, as well as various roles and responsibilities of auditors, training auditors and technical experts.

Frequently Asked Questions
- **Can internal audits at the Biobank be carried out by the Biobank employees?**
  Yes, but it is recommended that the Biobank employees undergo training for internal auditors. It is also important that the employee does not audit the area for which they are responsible.

References

ISO 20387:2018 Biotechnology – Biobanking – General Requirements for Biobanking.
ISO 9001:2015 Quality management systems—Requirements.
ISO 19011:2018 Guidelines for auditing management systems.

Improvement

13

Abstract

Biobanking is an activity that is influenced by many external and internal issues. The organization's response to changes in the scope of such issues is crucial for the proper implementation of processes and fulfillment of the requirements of recipients/users/customers. Improvement is understood as the process that allows for an early and effective adaptation to changes. It is a very broad issue and covers activities related to: (1) following the trends on the market, (2) introducing new products and services, (3) nonconformities and the resulting need to improve the QMS and Biobank processes in order to prevent recurrence, and (4) any feedback obtained. Improvement also allows the Biobank to meet changing legal and standardization requirements. This chapter describes the general requirements relating to the Biobank improvement process, including the periodic management review, which is a specific improvement tool.

- The Biobank shall develop, implement, and apply a process improvement procedure.

13.1 Tools for Improving the Effectiveness of Biobank

- The Biobank shall identify and use every possible tool for improvement.
- The Biobank shall seek feedback (including complaints) from suppliers, customers, donors, external institutions and organizations, and scientific partners. This information shall be analyzed with a view of improvement.

8.6.1, 8.6.2 ISO 20387:2018
9.1.3, 10.1, 10.3 ISO 9001:2015

The Most Common Practices

- Improvement may result from the review of procedures, audit results, corrective and remedial actions taken, management review and data analysis, risk and opportunity analysis, proficiency test results, taking into account the policy adopted and the objectives of the Biobank.

13.2   Management Review

- The Biobank shall analyze the data and review the organization's QMS at planned intervals to ensure its usefulness, adequacy, and effectiveness. This review shall cover the opportunities for improvement and the need for changes in QMS, including quality policy and quality objectives (see Sects. 2.2 and 2.3 MBQM).
- The Biobank shall implement a procedure for management review. The procedure shall describe the measures aimed at planning, preparation, conduct, and implementation of reviews.

8.9.1 ISO 20387:2018
9.3 ISO 9001:2015

13.2.1 Conducting a Management Review

- At least the Biobank Manager and key personnel shall participate in the management review.
- The person responsible for QMS in the Biobank shall identify the QMS areas where data will be collected and analyzed (creating the basis for system improvement in the Biobank).
- The management review shall be documented.
- The management review at the Biobank shall be carried out on the basis of the following documented input, containing the following information:
  - changes in identified internal and external factors affecting the functioning of the Biobank;
  - achievement of the objectives set;
  - the adequacy of the policies adopted;
  - the adequacy of the procedures in relation to the processes carried out;
  - the actions taken following the previous management review and their status;
  - the results of internal and external audits;
  - corrective and remedial actions taken;
  - the activities undertaken by the Biobank, including changes in them;
  - opinions (including complaints) obtained from customers, donors, external institutions and organizations, scientific partners;
  - the effectiveness of the improvement actions implemented and the possibilities of their development;

- the adequacy of the resources of biological material and data in the context of the needs of the Biobank and/or cooperating units;
- the results of the risk and opportunity analysis, and the actions taken following them;
- the results of quality control, process monitoring;
- the activities of external service and materials providers;
- other significant identified factors.

- The results of the management review shall be documented and include decisions as well as specify the planned actions in relation to:
  - quality management system;
  - processes;
  - activities related to meeting the requirements set for the Biobank (e.g., legal);
  - any identified opportunities for improvement;
  - the provision of resources necessary to achieve the Biobank's objectives;
  - the provision of the required biological material and data.

6.1.2, 8.9 ISO 20387:2018
9.3, 9.1.3 ISO 9001:2015

The Most Common Practices

- The management review is carried out in the form of a meeting of the Biobank Manager with key personnel, i.e. responsible for the functioning of the particular area.
- Employees responsible for the functioning of processes prepare data analysis on the functioning of processes in their areas and evaluation of the goals set in the previous management review.
- The management review at the Biobank is conducted on the basis of information on the functioning of processes (input data), including:
  - the number of publications published and the citation index obtained;
  - the number of implemented research projects;
  - the number of tests/analyses performed;
  - information on improving technological processes or implementing new ones;
  - the number of biological samples collected;
  - the list of new equipment, infrastructure development;
  - information on personnel (e.g. the number of trainings).
- The personnel conducting the management review are responsible for updating or setting new objectives in given areas.
- The results of the management review at the Biobank are used to take improvement actions, including measures aimed at:
  - changes in the number of processes (joining or separating existing processes);
  - starting new processes;
  - modification or decommissioning of existing processes;
  - changes in process documentation (the number or content of documents).

- The management review report contains information such as:
  - the date of the review;
  - the place where a meeting is held;
  - participants;
  - data analysis (data on the functioning of processes; significant changes in the QMS documentation; the results of audits; corrective actions taken; the review of risk and opportunity identified; feedback from customers, complaints; evaluation of the objectives set in the previous management review; setting new objectives);
  - recommendations for improvement;
  - the assessment of QMS in the Biobank.
- The management review report is signed by all meeting participants.

Frequently Asked Questions
- **Does the process improvement procedure have to be a separate document from the management review procedure?**
  No. The general procedure for improving the Biobank processes may include guidelines for conducting management reviews.
- **Can the management review be the only tool presenting the areas for improvement?**
  No. As indicated in Sect. 13.1, the management review is one of the tools that is used periodically, e.g., once a year. Improvement is a continuous process and should result from the processes mentioned in Sect. 13.1.
- **What is the review of identified risk and opportunity during the management review?**
  It consists in verifying whether there have been changes in previously identified risk and opportunity. Since the last assessment, there may have been a change in the factors causing this risk/opportunity, which may necessitate a change in operation that will lead to minimizing the risk or increasing the likelihood of using the opportunity.

References

ISO 20387:2018 Biotechnology—Biobanking—General Requirements for Biobanking.
ISO 9001:2015 Quality management systems—Requirements.

Biobank Cooperation in the Scientific, Research, and Development Area

14

Abstract

One of the primary objectives of science policy is to establish cooperation and conduct scientific research. The aim of scientific cooperation is to exchange knowledge in communities involved in the development of specific disciplines and to provide access to tools, research materials, and laboratories at the optimal level of research quality.

The information contained in this part of the book focuses on describing the rules of cooperation between the Biobank and other organizations, including scientific, research, development, and commercial cooperation. It discusses the issues related to the provision of biological material and data and the formalization of cooperation on the basis of an agreement, which should include the parties and subject matter of the agreement, requirements and conditions to ensure confidentiality, the detailed terms and conditions and scope of cooperation, as well as all necessary annexes resulting from the scope of cooperation. The issues of material and data transfer agreements were also addressed. Moreover, the chapter discusses the need to define the means of communication between the cooperating parties, including the obligation to notify each other of any deviations and irregularities.

14.1 General Requirements

- The Biobank shall establish, implement, and apply a procedure in the area of cooperation (scientific, R&D).

7.3, 7.6, Annex A—A.7 ISO 20387:2018

© Wroclaw Medical University 2023
A. Matera-Witkiewicz et al., *Manual of Biobank Quality Management*,
https://doi.org/10.1007/978-3-031-12559-1_14

14.2   Sharing of Biological Material and Data for Research

- The Biobank shall establish, implement, and apply a procedure in the area of sharing of biological material and data.

7.3.1.1, 7.3.3, 7.4 ISO 20387:2018

The Most Common Practices
- One example of how the principles of sharing biological material and/or data can be presented is a description, a diagram, or a table. Below an example of diagram is presented:
  - partner's inquiry to the Biobank;
  - the response of the Biobank with the information about the availability of biological material and/or data;
  - providing by the partner of a short description of the purpose for using biological material and/or data;
  - the decision of the Biobank to begin/reject the cooperation;
  - common request to the Bioethics Committee for research objectives approval;
  - signing the cooperation agreement;
  - implementation of the agreement.

14.3   Cooperation Agreements

- The cooperation agreement between the Biobank and a potential partner is an undertaking according to the internal regulations.
- The cooperation agreement shall include the following parts:
  - the parties and subject matter of a contract;
  - requirements and conditions to ensure the confidentiality;
  - the detailed terms, conditions, and scopes of cooperation;
  - all necessary annexes resulting from the scope of cooperation.

7.3.3, 7.6 ISO 20387:2018
8.2.1, 8.2.2, 8.2.3 ISO 9001:2015

The Most Common Practices
- Examples of contract annexes can be as follows:
  - the type and quantity of the material for testing;
  - the type and quantity of the material stored;
  - information on documentation provided with samples (if not included in the contract);
  - procedures/instructions for the collection of biological material, coding, transport, issue, a model informed consent for donors;
  - the conditions for handling biological material at the end of the contract;

- BBMRI-ERIC recommends its members to follow the recommendations from MTA (material transfer agreement) and DTA (data transfer agreement). Details can be found under the following links: http://www.bbmri-eric.eu/wp-content/uploads/DATA-TRANSFER-AGREEMENT.pdf
    http://www.bbmri-eric.eu/wp-content/uploads/MATERIAL-TRANSFER-AGREEMENT.pdf
- Such contracts, apart from the parties and the subject matter of a contract, include, e.g.: biological material and/or data disposer, sharing condition, confidentiality clauses, responsibilities, intellectual property.

14.4 Communication During Scientific Cooperation

- The Biobank shall define and record the rules of communication as part of scientific cooperation.
- Explanations and regulations shall be documented.

14.4.1 Informing About Irregularities in the Course of Research

- The Biobank and the Partner shall inform the second party about irregularities which occurred during cooperation.

6.4.1.4, 7.11.1.3, 7.11.1.5 e) ISO 20387:2018
8.2.4, 8.3.6, 8.5.3, 8.5.6, 9.1 ISO 9001:2015

Frequently Asked Questions
- **How do I inform my partner about irregularities?**
  It can be done by writing or by email. Please note that it should be a form provided in the documentation.
- **What kind of irregularities do I need to report?**
  These should be all irregularities that affect the safety and quality of the research being conducted: E.g.:
  - *storage conditions not in accordance with the contract (e.g., freezer failure and thawing of samples);*
  - *loss of some data (e.g., unintentional deletion of some data from the database);*
  - *unintentional use of out-of-date reagents for some of the analyses conducted.*

References

ISO 20387:2018 Biotechnology—Biobanking—General Requirements for Biobanking.
ISO 9001:2015 Quality management systems—Requirements.

Safety and Security

15

Abstract

According to the literature, the factor that influences the willingness to make donations to Biobanks is trust, namely donor's trust that his/her samples and related data will be used properly and stored in a secured manner. To ensure it, appropriate security mechanisms should be implemented. To address such challenges, this chapter outlines some basic threats and risks, as well as solutions and technical improvements that may mitigate them. Owing to their experience gained during information security audits and workshops, the authors draw attention to frequently neglected aspects which may decrease information security of an entity. The chapter highlights aspects related to safety procedures in the context of employees, biological material, and related data. Attempts were made to emphasize the importance of being aware of the functioning of the Biobank within the organization and the related responsibilities assigned to the head of the unit. The authors described basic methods of securing the IT infrastructure and paid attention to aspects related to access to Biobank resources—both in terms of physical access and data stored and processed in the Biobank. It was emphasized how important it is for the organization to have the appropriate approach to regular backups, especially in the case of using external services, including cloud data processing. Each subsection tries to convey the most common practices in relation to the most common questions.

15.1 Security Procedures

- The Biobank or a unit of which it is a part (the Biobank Manager or a person designated by the Manager) shall develop, implement, and apply a policy that minimizes the risk of negative impact on the health and safety of Biobank employees, researchers, and other persons working/staying in the Biobank premises.

- The Biobank or the unit of which it is a part (the Biobank Manager or a person designated by the Manager) shall develop, implement and apply a policy ensuring the safety of biological material and accompanying data/documentation.
- The Biobank or the unit of which it is part (the Biobank Manager or a person designated by the Manager) shall communicate the information:
  - to the Biobank employees concerning the persons responsible for:
    (a) their safety during the technological process by complying with the internal rules of personal protection (e.g., the use of laboratory coats, protective gloves);
    (b) the safety of biological material during the technological process by using laboratory devices in accordance with their intended use, the security of data related to biological material, and the data of donors/probants collected in the Biobank.
  - to the external entities cooperating with the Biobank concerning the persons responsible for:
    (a) the safety of the study participants during the collection of biological material;
    (b) the safety of persons collecting biological material;
    (c) biological material safety during transport;
    (d) the safety of persons transporting biological material—if applicable;
    (e) the safety of people cleaning the premises of the Biobank;
    (f) any event consisting in violating the security procedure shall be immediately reported to the Manager of the Biobank;
    (g) the Biobank shall register every notification of a security breach (see Sect. 1.4.1, Chaps. 4, 7, and 9 MBQM).

15.2   Biological Material Safety

- The Biobank or unit of which it is part shall provide (see Chaps. 8 and 10 MBQM):
  - the ability to monitor (track) biological material at every stage of the technological process;
  - the ability to monitor (track) the environmental conditions in which biological material is stored/processed at every stage of the technological process;
  - the ability to monitor (track) people who gain access to biological material at every stage of the technological process.

6.3 ISO 20387:2018

15.3 Information Security

- Due to the possible processing by the Biobank of the personal information collected, in particular concerning the donor, the Biobank Manager or a person designated by the Manager shall provide an adequate system to ensure the safety of the processed information, in particular:
  - information on stored/biobanked material;
  - Biobank management system documentation;
  - personal data of the personnel of the Biobank;
  - personal data of donors.

4.3, 6.2.1.2, 6.4, 7.3.2, 7.7.1, 8.5.1, 8.5.2 ISO 20387:2018
3.5.3 ISO 27000:2014
12.1.3 ISO 27002:2013

The Most Common Practices
- **Examples of risk areas to consider when conducting risk assessment**
  - **Unauthorized physical access to the premises**—it is necessary to analyze whether the rooms in which the data are processed or stored are properly secured. The chance of access to this type of premises by unauthorized persons should be assessed and the effects of obtaining such access should be determined. Risk minimization methods—controlling the access to rooms through: supervision of access to rooms, magnetic card system, access code system, key system.
  - **Unauthorized access to data/information**—the risk of unauthorized access to data by employees and external parties should be analyzed. The above risk may be considered in two variants—accidental and intentional unauthorized access. Risk minimization methods—control of access rights to data or applications, proper security of the IT infrastructure (basic security methods, described in Sect. 15.4, may prove insufficient—each selection of mechanisms and solutions should be preceded by a risk analysis). An appropriate IT system that meets the minimum requirements is described in Sect. 15.4.10 (the described requirements may prove to be insufficient—each time the choice of the IT system managing information in the Biobank should be preceded by the risk and business analysis).
  - **Violation of data integrity**—the risk of data integrity breaches should be analyzed. The analysis should be performed in terms of accidental and intended activities of both the personnel of the Biobank and external persons—see "Integrity—the property of ensuring accuracy and completeness" (PN-ISO/IEC 27000:2014). Risk minimization methods—the control of access rights to data or applications, proper IT infrastructure protection (basic security methods described in section "Basic Methods for IT Infrastructure and Processed Data Protection" may prove insufficient—each selection of mechanisms should be preceded by risk analysis), an appropriate IT system that meets the minimum requirements described in Sect. 15.4.10 (the

requirements described may prove to be insufficient—each selection of the IT system managing information in the Biobank should be preceded by the risk analysis and business analysis), checking the quality of data received.

- **Data loss**—analyzes the risk of data loss. The analysis should be carried out in terms of accidental deletion of data, intentional deletion of data, another form of data loss, e.g. ransomware attack, IT infrastructure failure. Risk minimization methods—the control of access rights to data or applications, proper security of IT infrastructure (the basic security methods described in Sect. 15.4 may prove insufficient—each selection of mechanisms should be preceded by the risk analysis), appropriate IT system meeting the minimum requirements described in Sect. 15.4.10 (the requirements described may prove insufficient—each selection of the information system managing information in the Biobank should be preceded by the risk assessment and business analysis), appropriate backup/recovery of backup copies, a configuration of storage space enabling redundant recording of information, e.g. distributed file systems or the use of RAID at the appropriate level (first or higher).
- **Loss of business continuity**—assesses the risk of business continuity loss. The analysis should be made in terms of the occurrence of phenomena that prevent the normal operation of the Biobank, e.g. data loss, IT infrastructure failure, power loss, flooding, etc. Risk minimization methods—policy/procedure for making backup copies (Sect. 15.4.11 presents the factors which are worth considering when planning a backup system), procedures for restoring business continuity.
- **Data leaks/breaches**—analyze the risk of data leakage/breach. The analysis should be performed in terms of data leakage from the organization caused by the intentional or accidental action of personnel, intentional, or accidental action of external partners, the possibility of an attack aimed at stealing or disclosing data. Risk minimization may involve the introduction of rules for the use of cryptographic tools, anonymization or pseudonymization of processed data, the control of access rights to data or applications, proper security of IT infrastructure (basic security methods described in Sect. 15.4 may prove insufficient—each selection of mechanisms should be preceded by the risk analysis).

Frequently Asked Questions
- **What is the information Security Management System?**
  The information security management system should be understood as an action aimed at minimizing the risk of data/information leakage, gaining access to information/data by unauthorized persons, the violation of information/data integrity, data loss, the loss of business continuity through the adequate selection of methods, resources, and regulations.
- **How to start implementing the Security Management System?**
  The basis for implementing the Security Management System is to conduct the risk analysis in the area of information security (recommendations for risk analysis are described in Sect. 2.4 MBQM).

- **What consequences can be associated with a data leak?**
  The most harmful cost of the leakage of data processed in Biobanks may be the loss of public confidence in entities such as Biobanks and biobanking itself. It can also be penalties imposed by regulators that can reach millions of euros.
- **How to protect the Biobank against unauthorized access?**
  Protection against unauthorized access to the premises—the system of authorizations issued by the Head of the Biobank or a person authorized by him—a person not having valid permissions cannot access the room (e.g., a magnetic card does not open the door lock, the employee will not be given the keys to the room, which he or she is not allowed to enter).
  Protection against unauthorized access to data can be obtained by implementing IT system management instructions and/or information security policy, specifying procedures for granting and withdrawing rights to data or IT resources.
- **How to protect the Biobank against data integrity violation?**
  Protection against data integrity violations may consist in implementing IT system management instructions and/or information security policy, specifying procedures for granting and withdrawing access rights to data or IT resources; using logging of changes introduced in data sets, designating curators of data sets.
- **How to protect the Biobank against data loss?**
  Protection against data loss may consist in implementing IT system management instructions and/or information security policy, determining the procedures for granting and withdrawing access to data or IT resources, policy/procedure for making backup copies (the factors that have been described in Sect. 15.4.11 are worth considering when planning a backup system).
- **What does integrity violation or data loss mean?**
  Examples of data breaches or data loss:
  An employee with appropriate permissions accidentally modifies incorrect fields in the IT system, e.g. regarding a donor. The action will not be blocked by the IT system due to the user's appropriate rights. The only possibility of detection consists in the reference to system logs.
  An employee with the appropriate permissions deletes a part of the data set—this action is allowed by the system's security mechanisms—the employee had the appropriate permissions. The only repair option is the data recovery from backup.
  The Biobank's computers have become the target of a ransomware attack—processed data has been encrypted—in such a situation the repair method will be the data recovery from backup.
- **How to protect the Biobank against the loss of business continuity?**
  Protection against the loss of business continuity may involve the use of a backup copy system protecting against data loss, the use of an emergency power supply system, prognosing the need for data storage space.

15.4   Basic Methods of Securing IT Infrastructure and Data Processing

15.4.1 Awareness of the Functioning of the Biobank in the Organization

- If the Biobank is part of a bigger entity like hospitals, universities, or scientific organization, the Biobank Manager or a person designated by the Manager shall inform the personnel responsible for managing information security about the functioning of the Biobank and about the emergence of new risks within the organization. As a rule, these comprise: Data Protection Officer and the representatives of IT services: Network Administrator or IT Systems Administrator (in different organizations they can be referred to differently). A good idea would be to perform with them a joint risk assessment, as a starting point for the selection of adequate security methods and measures.
- The Biobank operations shall comply with internal regulations on information security or the processing of personal data in force in the organization and in line with the applicable law.

15.4.2 General Principles for the Secure Exchange of Information

- Access to data can only be obtained by a person who has the appropriate authorization given by the Biobank Manager or a person designated by the Manager.
- Condition for granting the access shall be signing a confidentiality agreement.
- Biobanks data sets shall be cataloged and need to contain information about data curators for each of them.
- The personnel of the Biobank shall only use official business email to handle all business matters.
- Every staff member of the Biobank conducting correspondence shall have a dedicated email account. It is unacceptable to use accounts available for more than one user, it is also unacceptable to periodically transfer email account handling to other employees, e.g. during the holiday leave—this may hinder the identification of the sender/recipient of the information as well as the information that was sent in such a situation.
- If the use of portable memory is allowed in the Biobank, an obligation to use registered portable memory devices provided by the organization shall be introduced. The optimal method is to use IT solutions that prevent connecting devices other than registered portable memory devices to the workstation.
- The Biobank Manager or a person designated by the Manager shall specify the rules for using mobile devices such as laptops, tablets, telephones/smartphones. The security mechanisms shall be introduced. A decision shall be made whether staff may use private devices to carry out their duties or not. The above rules shall

be written in the form of a procedure or instructions and be clearly communicated to the personnel of the Biobank.

- Data output from the Biobank may take place only on the basis of specific procedures taking into account the type and nature of the data, the method and conditions of their transmission (see Chap. 5 MBQM), adequate methods of data security, e.g. by using cryptographic tools, the methods of confirming the receipt and quality of data by the recipient. These also include the data related to the liquidation of the Biobank and emergency situations—requiring business continuity.

15.4.3 Security of Processed Data

- Where possible, the Biobank shall implement the mechanisms for anonymization or pseudonymization of donor data in accordance with the requirements described in Sect. 5.5 MBQM.
- Unless the tasks performed require otherwise, the Biobank shall accept only anonymized data, and in the case of biological material, information about the donor shall be encoded in a way that does not reveal personal data or identification numbers (e.g., Personal ID).
- Processed paper documentation containing donor's personal data, e.g. the donor's consent to participate in the study, shall be stored in a locked cabinet or equivalent place (optimally in a safe) in a room with restricted access and adequately secured (see Sect. 15.4.4 MBQM).
- The rooms, infrastructure, and IT systems used in data processing shall be appropriately secured.

15.4.4 Basic Methods of Securing Access to the Premises

- Access to the premises where data processing takes place can only be obtained by authorized persons—authorization shall be issued by the Manager of the Biobank or a person designated by the Manager.
- There shall be a list of authorized personnel in the Biobank containing information on the rooms which a person is authorized to access.
- The list shall be regularly updated and forwarded to the relevant personnel responsible for the distribution of keys.
- Keys securing the access to the rooms shall not leave the building and shall be issued only to authorized personnel.
- In the case of securing access with cards or codes, both the card and the code shall be assigned to the unique employee.
- The card or access code assigned to the user shall provide access to the rooms resulting from the employee's authorization. It is allowed to grant access to common rooms, such as corridors or staircases, if it is necessary to exercise access rights.

- If workstations/servers used in the Biobank business processes are located in a dedicated server room, the rules for accessing that room shall also be specified.

15.4.5 Operating Systems and Software

- Workstations providing access to IT systems and storage space shall be equipped with operating systems with actual manufacturer support.
- Operating systems shall be kept up to date.
- Tool software installed on workstations providing access to IT systems and storage space shall have the current support of the software manufacturer.
- The software shall be updated on a regular basis.
- In the event that the tool software is not automatically updated by the manufacturer there shall be a clear identification of the person responsible for the update process and the process itself shall be properly documented.
- In exceptional situations, when the specialized hardware, software works only on unsupported versions of operating systems, it is necessary to ensure proper protection of such computers, e.g. by disconnecting from the local computer network, data transfer can then take place via dedicated data carriers. The use of such an exception shall be properly documented.

15.4.6 Anti-virus Software

- Anti-virus software shall be installed on the workstations.
- The installed software shall be in the current version and have an actual virus database.
- Workstations shall be regularly scanned, the frequency of scanning shall be determined on the basis of a risk assessment.

15.4.7 User and Password Management

- Access to operating systems and data processing software shall be protected with a password. Deviations from this rule are permitted, but they shall be documented and properly justified.
- Requirements regarding password strength (length and type of characters) and the frequency of changes shall be determined on the basis of risk assessment and shall be codified in a formal document, e.g. procedure or information security policy.
- Significant improvement of work related to user management can be achieved using centrally managed systems.
- The Biobank shall specify the procedure to be followed in the event of the employment of a new employee, the termination of the contract with the employee, the change of a job (see Chap. 4 MBQM).

15.4.8 LAN Network

- The structure of the LAN shall be documented—the documentation may be done at the level of the Biobank or the Organization of which the Biobank is a part.
- Physical methods of securing access to the network shall be selected on the basis of a risk assessment carried out so as to minimize the possibilities of using vulnerabilities.
- If the Biobank uses the organization's network infrastructure, the Biobank network shall be separate, e.g. through the use of VLANs.
- Network activity shall be monitored on an ongoing basis—IDS/IPS tools can be used for this—network monitoring can take place at the Biobank itself or at the level of the institution of which the Biobank is a part—hospital, university, etc.

The Most Common Practices
- The most common security methods used to secure the access to the network are MAC addresses filtering, authorization of users or workstations using the LDAP or RADIUS protocol.

15.4.9 Emergency Power Supply

- The devices ensuring the maintenance of IT infrastructure power supply—UPS— shall provide power to critical devices for the operation of the Biobank until the emergency power source is activated.
- In the absence of an emergency power supply, the UPS shall provide power for the time of data storage and safe shutdown of IT systems.
- The Biobank Manager or a person appointed by him shall determine, on the basis of a risk analysis, the elements of infrastructure requiring power maintenance.
- The elements of the emergency power supply infrastructure shall be subject to regular tests confirming their efficiency and, in the case of UPSs, the verification whether the current capacity/battery life is sufficient to support the specified infrastructure for the assumed time.

7.7.2 d) ISO 20387:2018

15.4.10 Basic Features of the Biobank IT System

- The system shall provide the opportunity to identify the person making changes to the system.
- The system shall record the type of changes introduced in the system preferably by the versioning of data.
- The system shall provide access control to the data processed in the system, e.g. by granting permissions.

- The system needs to ensure unambiguous identification of biological material and data in accordance with Chap. 8 MBQM.

15.4.11 Backups

- Backups shall be performed regularly.
- The scope of data subjected to this process and the frequency of performing shall be determined by the Biobank Manager or a person designated by the Manager on the basis of risk assessment and business analysis; the technical capabilities of the entity shall be taken into account. When planning the above, one shall answer the questions how long the unit's downtime is acceptable and the loss of what amount of data is acceptable. The frequency of test recovery of backup copies shall also be established, and the process shall be documented. The Biobank Manager or a person designated by the Manager shall clearly define persons responsible for making backup copies and persons responsible for supervising the process.
- After completing the backup process, the verification of the correctness of the copies and the ability to restore data from the copy shall be performed.
- The process of making copies shall be formalized and documented.
- The storage of backup copies shall be properly designed.

Frequently Asked Questions
- **What should be the frequency of making backup copies?**
  The backup copy is made only once a month, e.g. on the last day of the month. If the failure occurs on the penultimate day of the month (the day before the date of making the copy), it means that the data collected or produced within the whole month is lost—the last copy was made a month earlier.

  A backup copy is made only once a week, for example, on Sundays. If the failure occurs on a Saturday—the day before the date of making the copy—it means that the data collected or produced within the whole week is lost—the last copy was made a week ago.

  A backup copy is made once a day. The occurrence of a failure means the loss of data collected throughout the day—the last copy was made the day before.

  A backup copy is made every hour. The occurrence of a failure means the loss of data collected within one hour—the last copy was made an hour before.

  The cost of restoring business continuity will mean the cost of re-entering or generating of the data that has been lost. As the frequency of making copies increases, the cost of handling them is higher as well, involving greater staff commitment and perhaps the need to suspend the data-generating processes. When planning the backup process, all aspects should be considered and the decision based on the results of risk analysis and business process analysis should be made.
- **How often do you verify the backed-up information?**

The sample size and frequency of checking the security copy should be specified in the copy procedure. The basis for making such a decision should be the risk analysis and the value of the information being secured.
- **What are the best practices for backup storage?**
The most common method is the "3-2-1" formula. This means: 3 copies of data must be produced each time, using different storage technologies (cloud solutions, flash drive, hard drive, tape, etc.) with at least one copy stored in a different location.

15.4.12 External Services Including Cloud Services

- In the case of the decision to use IT infrastructure service by external contractors or to use external resources to process or store data (the so-called cloud services/ cloud computing), the following shall be noted:
  - does the contract clearly define the responsibilities and expectations of the parties to the contract?;
  - if you use access to computing power or storage space, it is necessary to include the requirements for the quality of services provided (the so-called SLAs). It consists in determining the quality and availability of the service;
  - when concluding a contract regarding the quality of service, you shall precisely specify the methods of measuring the availability of the service, define the availability of the service, determine the consequences of the failure to provide the service, e.g. contractual penalties;
  - a minimum level of security provided for services shall be specified;
  - indicators shall be defined on the basis of which the quality of the service provided will be examined.

Frequently Asked Questions
- **What does it mean to guarantee the quality of services provided (SLA)?**
SLA—Service Level Agreement defines quality, availability, responsibilities of the parties. It may relate to the service availability, is defined as a percentage and means the length of time during which the service will be available (Table 15.1).

15.4.13 Personnel

- An important element of the Biobank information security infrastructure is personnel, who shall constantly improve their knowledge and build an information security culture. The principles applying to training are described in Chap. 4 MBQM.

5.8, 6.2.1.2, 6.2.3, 6.4, 6.5.1, 7.3.3, 7.5.1, 7.5.3, 7.7.2, 7.7.4, 7.8.3, 7.10, 8.5.2 ISO 20387:2018
6.1.2, 6.2, 7, 8.3.3, 9, 11, 12.2, 12.3, 12.5, 13.1, 15.2 ISO 27002:2013

Table 15.1 The examples of service unavailability times for specific SLA levels in different time slots

Time slots	Percentage level of service availability as defined in the SLA				
	80%	90%	99%	99.90%	99.99%
Daily	4 h 48 m 0.0 s	2 h 24 m 0.0 s	14 m 24.0 s	1 m 26.4 s	8.6 s
Weekly	1 d 9 h 36 m 0.0 s	16 h 48 m 0.0 s	1 h 40 m 48.0 s	10 m 4.8 s	1 m 0.5 s
Monthly	6 d 2 h 5 m 49.2 s	3 d 1 h 2 m 54.6 s	7 h 18 m 17.5 s	43 m 12.0 s	4 m 23.0 s
Annually	73 d 1 h 9 m 50.4 s	36 d 12 h 34 m 55.2 s	3 d 15 h 39 m 29.5 s	9 h 45 m 36.0 s	52 m 35.7 s

A higher level of reliability means greater responsibility for the contractor, which translates into a higher cost of service

References

ISO 20387:2018 Biotechnology—Biobanking—General Requirements for Biobanking.
ISO 9001:2015 Quality management systems—Requirements.
ISO 27000:2014 Information technology—Security techniques—Information security management systems—Overview and vocabulary
ISO 27002:2013 Information technology—Security techniques—Code of practice for information security controls

Glossary

Accessioning (*logging*) The act of documenting the addition of new biological material or related data to the Biobank (based on ISO 20387: 2018, definition 3.1).

Acquisition The act of obtaining ownership of biological material and/or associated data by the Biobank (based on ISO 20387: 2018, definition 3.2) in a situation where the Biobank is not responsible for its sampling or an act of establishing custody of such material and/or data. An example of acquisition of biological material or related data may be its purchase by way of donation, by way of purchase, by way of exchange.

Advisory Board Is a body that provides non-binding strategic and scientific advice to the management of Biobank; Advisory Board(s) could cover scientific issues and ethical, legal, and societal issues (ELSI) as well.

Agreement Mutual acknowledgement of terms and conditions under which a working relationship is conducted (based on ISO/IEC TR 29110-5-1-3: 2017, definition 3.1); establishing mutual rights and obligations to perform research and/or scientific work.

Annex/appendix Forms, prints, labels, algorithms, hardware, cards, instructions, and partnership agreements referred to by a given procedure, instruction, or regulations.

Anonymization A process that removes association between the identifying data set and the data subject (based on ISO/TS 17975: 2015, definition 3.1.) Anonymization prevents the identification of a donor, its effects are irreversible or its reversal requires a disproportionate amount of time, effort, and costs.

Archival [copy] This is related to the storage of data over a prolonged period (based on ISO/TS 27790: 2009, definition 3.6). It does not require continuous access and there may not be another copy of these data.

Associated data Any information affiliated with biological material including but not limited to research, phenotypic, clinical, epidemiologic, and procedural data (based on ISO 20387: 2018, definition 3.3).

Audit A systematic, independent, and documented process for obtaining objective evidence and evaluating it objectively to determine the extent to which the audit criteria are fulfilled. (based on ISO 19011: 2018, definition 3.1). There are several types of audit: an external audit—referred to as a second- and third-party audit), an internal audit (referred to as a first-party audit).

Auditor A person who conducts an audit [based on ISO 19011: 2018, definition 3.15] responsible for planning and efficient and effective execution of entrusted tasks, documenting observations from the audit.

Audit team One or more persons conducting an audit, supported if needed by technical experts (based on ISO 19011: 2018, definition 3.14). The audit team may consist of: a *lead auditor*—the auditor responsible for all audit phases, who selects and supervises the members of the audit team, prepares an audit plan, conducts opening, review, and closing meetings; a *Technical expert*—a person with specialist knowledge or skills, who provides specific knowledge or expertise to the audit team (based on ISO 19011:2018, definition 3.16), and an O*bserver*— individual who accompanies the audit team but does not act as an auditor (based on ISO 19011:2018, definition 3.17). An A*uditor in training* can also be included in the audit team, but should participate in the audit under the direction and guidance of the designated auditor.

Backup [copy] A duplicate of stored data (based on ISO 18943: 2014, definition 2.2). Backups are short-term in nature. They are used to reconstruct the original data in case of loss or damage.

Biobank A legal entity or part of legal entity that performs biobanking (based on ISO 20387: 2018, definition 3.5).

Biobanking A process of acquisitioning biological material and/or associated data through its collection or purchase and the process of its storage, along with some or all of the operations related to its processing, preservation, distribution, and provision of samples and data, as well as testing, analysis, and publishing (based on ISO 20387: 2018, definition 3.6). The definition does not apply to reproductive cells, gonads, embryonic or fetal tissues, and reproductive organs or parts thereof.

Biological material Any material from a biological unit entity such as a human, animal, plant, microorganism, or multicellular organism(s) that is (are) neither animal nor plant or any substance derived from such an entity (based on ISO 20387: 2018, definition 3.7).

Biosafety Principles of limitations of spread as well as technologies and practices used to prevent unintended exposure to pathogens and toxins or their accidental release (based on ISO 20387: 2018, definition 3.8).

Biosecurity Institutional and personal security measures and procedures designed to prevent the loss, theft, misuse, improper use, or unwanted release of pathogens, genetically modified organisms, toxin-producing organisms or parts thereof as well as toxins that are stored, transferred, and/or provided by Biobank (based on ISO 20387: 2018, definition 3.9).

Broad consent Consent for an unspecified range of future research subject to a few content and/or process restrictions. Broad consent is less specific than consent for each use (specific/strict/restricted consent), but more narrow than open-ended permission without any limitations ("blanket" consent).

Calibration/adjustment An action that, under certain conditions, first: defines interdependence between calibrated values set with a calibration template, together with their uncertainties of measurement, and equivalent indications, together with their uncertainties of measurement; and second: uses the data to define interdependence allowing measurement result to be obtained on the basis of indication. Calibration/adjustment is a set of operations that make it possible to define, under certain conditions, a relation between values measured with a measuring instrument or measuring system or values represented with a measurement template or reference material, and the right values achieved with measurement unit templates.

Certification/Legalization Third-party attestation related to equipment, products, processes, systems, or persons (based on ISO/IEC 17000: 2004, definition 5.5).

Clinical purpose The procurement and use of human cells, tissues, and organs for prophylactic, diagnostic, or therapeutic purposes in the provision of healthcare.

Coding/tagging Labeling of biological material for the purpose of identification, location or to provide other information (based on ISO 20387:2018, definition 3.48).

Collection (*procurement*) Activities, as a result of which biological material is obtained from the donor for the purpose of biobanking. In the case of humans, two scenarios for the collection of biological material are distinguished. In the first scenario, the acquisition of cells, tissues, organs or parts thereof, body fluids, etc., for biobanking is associated with their direct removal from the human system (intentional intake), while in the second—biological material obtained is cells, tissues, organs or parts thereof, body fluids, etc., that have been removed from the human system for reasons other than their collection for biobanking (non-intentional collection, acquisition/protection of medical waste material). An example of intentional collection is to take urine directly from the bladder through a suprapubic puncture or from a catheter inserted for this purpose. An example of nonintentional collection is urine collection during urination or through a pre-inserted catheter. In addition, the collection of biological material can be instrumental (it is necessary to use equipment such as needle, catheter or tools such as an endoscope) or non-instrumental (without the use of equipment, tools), e.g. sampling urine into a container during urination. In addition, collection may be invasive if related to the discontinuity of the shells of the body or the use of an endoscope or it can be non-invasive.

Competence The ability to apply knowledge, experience, and skills to achieve intended results (based on ISO 20387: 2018, definition 3.13).

Consent Freely given agreements based on adequate information obtained prior to the collection/use of participant's data (based on ISO 20252: 2019, definition 3.20).

Contract A binding agreement (based on ISO 9000: 2015, definition 3.4.7.).

Correction An action to eliminate a detected nonconformity. A correction can be made in advance of, in conjunction with or after a corrective action (based on ISO 9000: 2015, definition 3.12.3).

Corrective actions Actions to eliminate the cause of a nonconformity and to prevent its recurrence. A correction can be performed before, together with, or after a corrective action (based on ISO 9000: 2015, definition 3.12.2).

Critical Having a potential effect on the fitness of biological material and/or associated data for the intended purpose (based on ISO 20387: 2018, definition 3.16).

Critical equipment Equipment that is necessary to carry out the assumed process.

Critical process A process relevant to the operation of the Biobank, which, if disrupted, may have a serious impact on its operations.

Defect A nonconformity related to the intended or specified use (based on ISO 9000: 2015, definition 3.6.10). It may refer to equipment, services.

Disposal The act of removing of biological material and/or associated data from the Biobank, usually for scrapping, irreversible destruction, or returning to the provider/donor (based on ISO 20387: 2018, definition 3.19).

Document Information and the medium on which it is contained (based on ISO 9000: 2015, definition 3.8.5). An example of a document is a record, specification; an example of a carrier is paper, computer disk, mobile equipment containing non-volatile flash memory (memory stick).

Documentation register A list of the QMS documentation, which records the date of the first version of the document, information about the currently valid version, annexes, the founding document, and the persons responsible for implementation.

Documented information Information required to be controlled and maintained by an organization and the medium on which it is contained (based on ISO 9000: 2015, definition 3.8.6). Documented information may refer to: the management system, including related processes; information created in terms of the organization's activities (documentation); proof of achieved results (records). Documented information can be used to communicate, transfer messages, share knowledge, and provide evidence for what has been planned and actually done. It can occur in any form such as paper or electronic and on any type of media and can derive from any source. One of the sources of documented information are records from documenting the activities carried out in the organization.

Donor A living or dead biological unit such as a human being, an animal, a plant from which biological material and/or associated data is/are derived, which has/have been collected for biobanking (based on ISO 20387: 2018, definition 3.22).

Dynamic consent A consent model in which the participant can give/change his preferences over a longer period of time, e.g., by obtaining automatic electronic information about projects in which his/her biological material and/or data are used (based on *Pawlikowski J., Krekora-Zając D., Marciniak B.: Code of Conduct on the processing of personal data for the purpose of scientific research by biobanks in Poland (draft), 2019*).

Exception A deviation from the approved instruction or established requirement.

Improvement An activity to enhance performance (based on ISO 9000: 2015, definition 3.3.1). Optimization of processes, including the QMS, in order to improve specific actions. It should be systematically planned, initiated, and managed by a person responsible for a given process. Improvement is part of quality management aimed at increasing the ability to meet quality requirements.

Information Significant data (based on ISO 9000: 2015, definition 3.8.2). See documented information. It can be understood as:– A Biobank accreditation tool– A source of information about the internal situation at the Biobank— diagnosis and correcting potential discrepancies at an early stage– A tool for continuous improvement of the system and processes– A tool for continuous monitoring of the quality system and processes at the Biobank– A tool for continuous monitoring of the quality system and processes at the supplier/ subcontractor– A tool for mutual understanding in terms of quality requirements (affecting the high quality of biological material)– A tool for obtaining a certificate for a selected system– A tool for obtaining data allowing suppliers/ subcontractors to be selected and classified– A tool to help improve the quality system of suppliers/subcontractors– A tool to improve the functioning of the Biobank– A tool to increase the confidence of current and potential customers and associates of the Biobank

Inspection Determination of conformity with specified requirements (based on ISO 9000: 2005, definition 3.11.7); conformity evaluation by observation and judgement accompanied as appropriate by measurement, testing, or gauging. A periodic check of the technical condition of equipment for the fulfillment of certain requirements; it should be documented with adequate records.

Installation qualification (IQ) A process of establishing by objective evidence that all key aspects of the process equipment and ancillary system installation comply with the approved specification (based on ISO 11139: 2018, definition 3.220.2).

Life cycle Consecutive and interlinked processes applied to biological material and related data from collection, if applicable, acquisition, or reception to distribution, disposal, or destruction (based on ISO 20387: 2018, definition 3.29).

Main process A process constituting the operational basis of the Biobank, thus directly related to the goals of its functioning. Examples of such processes include the collection of biological material, its storage, processing (e.g., isolation of PBMCs), and distribution as well as scientific cooperation.

Material Any consumable material, including a reagent, used in technological processes in the Biobank.

Non-conforming product/service Output that does not meet requirements, e.g. biological material that does not meet the specification required by the recipient.

Nonconformity Non-fulfillment of the requirement (based on ISO 9000: 2015, definition 3.6.9). Failure to meet particular requirements (a product/manufacture/service that does not meet the accepted requirements).

Occupational exposure Exposure of employees during routine operations to blood and other potentially infectious materials, which is associated with the possibility of an infection. Exposure is understood as a breach of the skin's continuity through stabbing, scratching, cutting with a tool contaminated with infectious material; contact of infectious material with damaged skin (open wounds, cuts, cracks, scratches, abrasions of epidermis); splashes of infectious material over mucous membranes, causing conjunctivitis; long-term contact of undamaged skin with a large volume of infectious material.

Operational qualification (OQ) A process of obtaining and documenting evidence that installed equipment operates within predetermined limits when used in accordance with its operational procedures (based on ISO 11139: 2018 definition 3.220.3).

Performance qualification (PQ) A process of establishing by objective evidence that the process, under anticipated conditions, consistently produces a product that meets all predetermined requirements (based on ISO 11139: 2018, definition 3.220.4).

Preventive action An action to eliminate the cause of a potential nonconformity or other potential undesirable situation (based on ISO 9000: 2015 definition 3.12.1).

Procedure A specified way to carry out an activity or a process (based on ISO 20387: 2018, definition 3.35).

Process A set of interrelated or interacting activities that use inputs to deliver an intended result (based on ISO 9000: 2015, definition 3.4.1). Biobanks distinguish two types of processes (first being processes directly related to the handling of biological material and related data; the other being supporting processes). Depending on the reference context, the term "intended result" is referred to either generally as "output (output data)" or more specifically "product" (e.g., a frozen sample of biological material) or "service" (e.g., distribution of a frozen sample of biological material). Process inputs are usually outputs of other processes, and process outputs are usually inputs of other processes.

Provider (*depositor*) A person or entity from whom/which biological material and/or associated data are received or acquired for biobanking (based on ISO 20387: 2018, definition 3.41).

Purchase The action of buying something (biological samples, data, research results).

Research Ethics Committee (REC)/Bioethic Committee (BC) An independent body that reviews and controls medical experiments and human subject research to be conducted in accordance with national and international law.

Requalification Periodic control of correct operation of the device.

QMS documentation Created documentation, which is used and sustained within functioning quality management system.

Qualification Documented verification if the equipment is suitable for the assumed process or purpose; we make a distinction between installation, operational, and process qualification.

Qualification process A process confirming the fulfillment of accepted eligibility criteria or disqualification/rejection when these criteria are not met.

Qualification process A process that covers all verification activities including all the items of the product (component, equipment, subsystem, and system) (based on ISO 10795: 2019, definition 3.185). Activities carried out to verify equipment based on the assumed parameters in the course of IQ, OQ, and PQ.

Quarantine/grace period Temporary isolation of the object, to avoid its usage. Quarantine is applied until the objective results of specific requirements will be achieved.

Reception A process of receiving biological material and/or data related to it supplied to the Biobank. Typically, it is assumed that the receipt of material and/or data is directly related to their subsequent verification based on pre-defined acceptance criteria. The result of verification is a decision regarding the acceptance of material or data received or the lack thereof. Unaccepted material or data are most often removed by returning to the supplier or disposal. Such reception is sometimes associated with the acceptance of biological material and/or related data by the Biobank.

Record A document stating the results achieved or providing evidence of activities performed (based on ISO 9000: 2015, definition 3.8.10). Records are usually not subject to change. Examples of entries are completed biobanking reports.

Rejected material An expired, defective, or damaged reagent or disposable material (e.g., a syringe). The Biobank can be informed by the manufacturer about a defect, e.g. single-use materials. A single-use material or reagent may be damaged during transport to or storage at the Biobank.

Remedial actions All activities, including limiting or temporary activities, undertaken to restore the initial or assumed state.

Requirement A need or expectation that is stated, generally implied or obligatory (based on ISO 9000: 2015, definition 3.6.4). Requirements may be set by various interested parties or by the organization itself (the Biobank in this case).

Sampling Taking samples for subsequent analysis (based on ISO 20387: 2018, definition 3.40). Scheduled material collection (including biological material) used for acceptance testing of the product/service. It also concerns documented information, i.e. sampling of records during the audit process.

Scaling A comparison of the scale of a measured value with the template of a given scale. Scaling may be performed internally at the Biobank with the use of the purchased template scale that is adequately certified.

Specification A document that indicates requirements (based on ISO 9000: 2015, definition 3.8.7).

Stakeholder A person or organization that can affect, be affected by, or perceive itself to be affected by a decision or activity (based on ISO 9000: 2015, definition 3.2.3)

Supporting process A process supporting (supplementing) a technological process, e.g. a process regarding the employment of an employee at the Biobank.

Technological process A planned set of interrelated activities that result in the intended product/intended service. The technological process may change the initial properties of the material (physical, chemical) in order to obtain the established product/service.

The worst-case scenario A condition or set of conditions including upper and lower process parameters and circumstances which, under standard operating procedures, present the greatest risk of failure for a material or process compared to ideal conditions, which do not necessarily lead to defective material or an erroneous process.

Tiered consent A consent model in which the subject receives a set of options allowing him/her to indicate to what extent his/her biological material and data may or may not be used, e.g., with the exception of specific types of studies (based on *Pawlikowski J., Krekora-Zając D., Marciniak B.: Code of Conduct on the processing of personal data for the purpose of scientific research by biobanks in Poland (draft), 2019.*).

Top management Is the person or group of people who, at the highest level, lead and supervise the organization; in the case of a Biobank, this may be both the head of the Biobank and the director/manager/head of the unit in whose structures the Biobank is located. It is also permissible to delegate the responsibilities of top management to a designated employee of the entity; this depends on how the Biobank defines the authority (based on ISO 9000:2015, definition 3.1.1).

Traceability The ability to trace the history, application, or location of an object (based on ISO 9000: 2015, definition 3.6.13).

Validation Confirmation, through the provision of objective evidence, that the requirements for the specific intended use or application have been fulfilled (based on ISO 20387: 2018, definition 3.52). An action that aims to confirm or document (prove) that a process performed within the set range of parameters is performed effectively and in a repeatable way, and fulfills the set requirements of a specification and qualitative criteria.

Verification Confirmation, through the provision of objective evidence, that specified requirements have been fulfilled (based on ISO 9000: 2015, definition 3.8.12). With regard to equipment, it refers to an action associated with the measuring equipment that aims to confirm that the measuring equipment, during usage and between the set calibrations/adjustments, fulfills the requirements defined by the user.

Verification of QMS documentation Periodic review of documents, updates, and re-approval.

Printed in the United States
by Baker & Taylor Publisher Services